Quantitative Applications in

A SAGE PUBLICATIONS SERIES

Quantitative Applications in the Social Sciences

A SAGE PUBLICATIONS SERIES

Series/Number 07-165

FRACTAL ANALYSIS

Clifford T. Brown
Larry S. Liebovitch
Florida Atlantic University

Los Angeles | London | New Delhi
Singapore | Washington DC

For information:

SAGE Publications, Inc.
2455 Teller Road
Thousand Oaks,
   California 91320
E-mail: order@sagepub.com

SAGE Publications India Pvt. Ltd.
B 1/I 1 Mohan Cooperative
   Industrial Area
Mathura Road, New Delhi 110 044
India

SAGE Publications Ltd.
1 Oliver's Yard
55 City Road, London,
   EC1Y 1SP
United Kingdom

SAGE Publications Asia-Pacific Pte. Ltd.
33 Pekin Street #02-01
Far East Square
Singapore 048763

Printed in the United States of America

Library of Congress Cataloging-in-Publication Data

Brown, Clifford (Clifford T.)
Fractal analysis/Clifford Brown, Larry Liebovitch.
    p. cm.—(Quantitative applications in the social sciences; 165)
Includes bibliographical references and index.
ISBN 978-1-4129-7165-2 (pbk.)
    1. Social sciences—Mathematical models. 2. Social sciences—Statistical methods. 3. Fractals. I. Liebovitch, Larry S. II. Title.

H61.25.B75 2010
514′.742—dc22   2009045118

This book is printed on acid-free paper.

10   11   12   13   14   10   9   8   7   6   5   4   3   2   1

Acquisitions Editor:	Vicki Knight
Associate Editor:	Lauren Habib
Editorial Assistant:	Ashley Dodd
Production Editor:	Brittany Bauhaus
Copy Editor:	Liann Lech
Typesetter:	C&M Digitals (P) Ltd.
Proofreader:	Jenifer Kooiman
Indexer:	Diggs Publication Services, Inc.
Cover Designer:	Candice Harman
Marketing Manager:	Stephanie Adams

CONTENTS

ABOUT THE AUTHORS

Clifford T. Brown, PhD, earned a BA in archaeology (cum laude) from Yale University. His master's and doctoral degrees, both in anthropology, are from Tulane University. He has published a number of articles on fractal analysis in archaeology and anthropology. He teaches undergraduate and graduate classes in the Department of Anthropology at Florida Atlantic University. Among the graduate classes is a required seminar titled "Quantitative Reasoning in Anthropology" that includes extensive statistical content for social scientists.

Larry S. Liebovitch, PhD, earned a BS in physics summa cum laude from the City College of New York and his AM and PhD in astronomy from Harvard University. He has published articles on fractal and nonlinear analysis in biophysical, medical, and psychological systems. He teaches both undergraduate and graduate classes in psychology, interdisciplinary science, and mathematics at Florida Atlantic University. Among his undergraduate courses are Applications of Fractals to Psychology for psychology students.

SERIES EDITOR'S INTRODUCTION

In a 1999 article in *Scientific American* titled "A Multifractal Walk Down Wall Street," Benoit Mandelbrot, the developer of fractal mathematics and a champion of its application to a variety of fields, argued for the relevance of fractals to understanding market volatility. Claiming that widely applied financial models meant to assess investment risks ignore the possibility of extreme market events, Mandelbrot went on to show how a fractal analysis can more realistically reproduce observed patterns of market fluctuation. As I write this introduction in late 2009, after a global market meltdown at least partly attributable to unrealistic assessment of risk, Mandelbrot's analysis seems prophetic.

Clifford Brown and Larry Liebovitch's monograph on *Fractal Analysis* is a compelling and wide-ranging treatment of the mathematics of fractals and their application to the social sciences. After a general introduction to the subject, Brown and Liebovitch explain how fractals can be used in the analysis of univariate distributions; describe two-dimensional fractal patterns; explain how various social processes can generate data with fractal characteristics; and take up a variety of more advanced topics, such as multifractals (used, for example, in Mandelbrot's analysis of markets).

The mathematical literature on fractals can be daunting, but Brown and Liebovitch cut through the difficulty of the subject and present it in a manner suitable for a broad readership. Because of its accessibility and lucidity, I expect that their monograph will have a significant impact on how social scientists think about data, and will result in the increased use of fractal analysis in the study of social phenomena.

Editor's note: This monograph was begun under the direction of the previous editor of the QASS series, Tim Futing Liao.

—*John Fox*
Series Editor

PREFACE

Mathematician Benoit Mandelbrot coined the word *fractal* to denote objects that recursively fragment into ever smaller pieces that bear some likeness to the whole object such that they form intrinsically rough and complex patterns. These kinds of patterns abound in nature, and so natural scientists have found fractal mathematics indispensable for investigating a wide array of phenomena. Fractal patterns appear in the social sciences as well. This book is a primer on the application of fractal mathematics and statistics to social phenomena.

Several early applications of fractal mathematics took place in the social sciences, such as Vilfredo Pareto's (1897) study of the distribution of wealth, Lewis Fry Richardson's (1948; 1960, but published posthumously) study of the intensity of wars, and George Zipf's (1949) studies of the distributions of word frequencies and city sizes. These ideas were known to specialists in their fields, but they remained isolated, quirky concepts until Mandelbrot created the unifying idea of fractals in the 1970s and 1980s. Since then, however, despite the recognition that Zipf and Pareto distributions represent fractal distributions, the social sciences have lagged behind the physical and natural sciences in applying fractal mathematics.

In recent years, however, the number of applications of fractal analysis in the social sciences has ballooned. Their variety has expanded as rapidly as their numbers. For instance, fractal analysis had been used by criminologists to study the timing of calls for assistance to police (Verma, 1998), by sociologists for the study of gender divisions in the labor force (Abbott, 2001), and by actuaries for the study of disasters (Englehardt, 2002). The surprising range of fractal phenomena in the social sciences calls out for a comprehensive survey that would investigate the commonalities among them, which could lead to a broader understanding of their causes and occurrence.

Purpose of the Work

We wrote this book with three goals in mind, one pedagogical, one persua-sive, and one substantive. Our chief aim was pedagogical: to offer a special-ized presentation of fractal analysis oriented to the social scientist. This book is the first broadly accessible introduction to fractal analysis for social scientists. It is a primer that uses straightforward, informal language to take the reader from elementary principles through actual analyses. It supplies the reader with all the tools necessary to attempt his or her own analyses. We have made no assumptions about potential readers' mathematical sophistication. We start with basic ideas like sets and distributions so that any undergraduate can follow the presentation. Basic concepts of calculus are referred to in places, but calculus is not required to understand most of the material. To cater to the widest possible audience, we have written in a simple, natural style. The informality of the style should not be mistaken for intellectual casualness. Rather, we wish the book to appeal to the widest possible range of readers, including not only senior researchers but also undergraduates and graduate students.

Of course, one could learn about fractals, as we did, by studying the abun-dant albeit challenging literature available in mathematics, physics, geology, and other such fields. And we have included references to such works for readers seeking greater detail than we could provide in this format. Thus, we have assembled in one text a wide range of useful, practical information about the techniques and applications of fractal analysis that we felt, from our experience as practitioners, would be most helpful for our audience.

Our second goal was to persuade those social scientists and students who are not familiar with fractal analysis that it may be relevant to their own research. We decided that the best way to achieve this goal was to present carefully selected examples, each of which typifies a range of similarly struc-tured problems. We chose the examples from a wide range of social sciences in the hope that every reader will find at least some ideas related to his or her particular concerns. Selecting examples was difficult because there are many from which to choose, and it was challenging to decide which ones would resonate for the largest number of readers. As an aside, our view of the social sciences is fairly broad. We would include not only the obvious disciplines, such as sociology, anthropology, and political science, but also economics, geography, criminology, archaeology, linguistics, education, and even law and history. We have touched upon many of these subjects, and we hope that at least one example or observation impels you to look in a new way at a particular problem, phenomenon, or data set in your field.

Our final goal was to document, albeit selectively, the range of fractal phe-nomena currently known in the social sciences, to encourage social scientists to look for commonalities among them. So, although this is not a comprehensive survey, we found the range and diversity of fractal social phenomena both curious and exciting, and we have tried to share that with you.

Organization of the Volume

We hope that the order in which the ideas are presented in the book seems natural and logical to the point of inevitability, but of course, initially, the organization was by no means self-evident, and the final structure embodies a series of difficult choices. As our chief goal was didactic, we gave priority to the clear and systematic exposition of fractal math for people with a background in the social sciences. Therefore, the resulting organization is different from that found in other introductory works on fractals, which are directed toward very different audiences.

We divided the material into six chapters. Chapter 1 introduces fractals conceptually, explains what they are, describes their characteristics, and parses their formal definition. In Chapter 2, we focus in detail on the analysis of fractal frequency distributions, and in Chapter 3, we explain how to analyze fractal patterns embedded in two dimensions. In other books on fractals, these two topics are usually presented in the opposite order: The analysis of geometrical constructions is placed before the analysis of frequency distributions. Why? Fractals embedded in two dimensions have an immediate visual appeal, and their fractal character can be described in a way that is highly intuitive and easy to understand. We reversed the order, however, on the theory that most social scientists have more training in statistics than geometry and are therefore familiar with frequency distributions. Thus, we thought that by starting out on familiar ground, we could navigate more easily to the suspect terrain of geometry.

Chapter 4 addresses how fractal patterns form in social phenomena. As social scientists, we are not usually satisfied with describing a pattern statistically or geometrically. We also want to know how the pattern grew or emerged. A number of dynamical processes have been proposed to explain the development of fractal patterns in society and culture, and we discuss a selection of them.

In Chapter 5, we take up more advanced topics in fractal analysis, including fractals embedded in three dimensions, self-affine fractals, fractal time series, multifractals, and lacunarity. In a volume of this scope, we could not go into great depth of detail on any of these topics, but we did wish to introduce them so the interested reader could acquire at least a conceptual sense of the diversity of fractal phenomena and the methods available for their analysis.

Chapter 6 is a brief conclusion in which we provide practical advice about fractal analysis. The appendix, available on Professor Liebovitch's Web site, provides software for fractal analysis. Finally, the reference section is more extensive than is customary in monographs like this, but we wanted the references to serve as a resource for readers who wished to explore the subject in more detail. Thus, we cited a number of relevant studies of fractal social phenomena, as well as a variety of technical works on fractal analysis that readers can consult for more information about the mathematical procedures covered here.

In sum, this is a practical introduction to fractal analysis in the social sciences that should be accessible to almost everyone. We develop the key concepts gradually and systematically. We describe social processes that produce fractals, and we introduce more advanced topics in fractal analysis. Finally, as will become apparent, we enjoy fractal analysis. We find it parsimonious, elegant, and even beautiful at times. We hope this book infects you with our excitement about fractals.

Acknowledgments

The authors and Sage gratefully acknowledge the contributions of the following reviewers:

Michael Biggs, *University of Oxford*
Courtney Brown, *Emory University*

List of Figures

Chapter 1

Chapter 2

Chapter 3

Chapter 5

List of Tables

To Flora and Rita

CHAPTER 1. INTRODUCTION TO FRACTALS

Fractal geometry measures roughness intrinsically. Hence it marks the beginning of a quantitative theory specific to roughness in all it[s] manifestations. Roughness is ubiquitous in nature and culture....This is why fractality is also ubiquitous and why fractal geometry will never lack problems to deal with. (Mandelbrot, 2005, p. 193)

Do you suffer from rough data? Are your data strongly nonlinear? Non-normal? Irregular? Do they display complex patterns that seem to defy conventional statistical analysis? If so, fractal analysis might be the answer to your problem.

Social scientists are accustomed to studying highly complex data sets; it comes with the territory. Our data frequently seem to resist analysis by traditional statistical methods. We are often delighted to identify statistically significant patterns in our data even when those patterns account for only a small fraction of the variance in the data. Indeed, the traditional popularity of factor analysis and its cousins speaks to the difficulty of interpreting our data because they are approaches that seek to discover and highlight the patterns in complicated data sets.

Social science data not only are complicated, but often fail to follow convenient distributions such as the normal, Poisson, or binomial distributions. Our data are often highly irregular and/or highly skewed. The nature of our data accounts for the popularity of nonparametric statistical tests in the social sciences. The non-normality and irregularity of so much of our data apparently derive from the complexity of social dynamics.

Fractal analysis provides an approach to the analysis of many of our awkward data sets. More important, it provides a rational and parsimonious explanation for the irregularity and complexity of our data. It turns out that our data are not behaving badly; they are obeying unexpected but common rules of which we were unaware.

Mandelbrot

Mathematician Benoit Mandelbrot (1967, 1983) is the father of modern fractal analysis. He was born in Warsaw, Poland, to Lithuanian parents, but the family moved to France before World War II. He displayed a precocious aptitude for mathematics, with a particular predilection for geometry. He studied math in France, where his uncle, Szolem Mandelbrojt, a student of Hadamard, was a distinguished mathematician at the Collège de France. Much of

1

Mandelbrot's original work was done while he was a researcher at IBM in the United States in the 1960s and 1970s. He subsequently held an appointment in the Mathematics Department at Yale University for many years and is now Sterling Professor Emeritus of Mathematical Sciences there.

Mandelbrot evidently possessed a restless mind and eclectic interests. He worked on a broad array of apparently unrelated problems in economics, linguistics, and engineering. But it was just these intellectual wanderings that led him to his most famous insight: that a multitude of diverse phenomena in these and other fields all exhibited similar patterns that he would come to call fractals. A few of these patterns had been known since the 19th century, but they were regarded merely as weird curiosities. Mandelbrot realized that these patterns were common, even ubiquitous, and were significant in many fields of inquiry. He saw that they all belonged to a single type of universal phenomenon that was frequently misunderstood and incorrectly analyzed. He set out to study them as a unit, so he coined the term *fractal* and explored the underlying nature of these patterns and the dynamical processes that create them. He succeeded in showing that many complex and irregular patterns traditionally believed to be random, bizarre, or too complex to describe are, in fact, strongly patterned and can be described by fairly simple algorithms.

What Are Fractals?

What are these patterns called fractals? We discuss Mandelbrot's formal definition of fractals below, but let's begin with an informal discussion. Fractal geometry is the study of the form and structure of rough and irregular phenomena. Fractals are *sets* defined by the three related principles of *self-similarity, scale invariance,* and *power law relations.* When these principles converge, fractal patterns form. The essential characteristics of fractal patterns are captured by a statistic called the *fractal dimension.* Much empirical work in fractal analysis focuses on (a) showing that a particular data set has fractal characteristics, and (b) estimating the fractal dimension of the data set. There are various techniques for carrying out these two tasks, and most of this book is devoted to the exposition of them. In the following pages, we explain the basic concepts introduced in this paragraph.

Sets

Fractals are defined as sets. It is no accident that Georg Cantor both founded set theory and also devised the first, primordial fractal set. A set, of course, is just a list or enumeration of elements, objects, or events. For our purposes, the significant fact is that almost any kind of data set can be fractal: points, lines, surfaces, multidimensional data, and time series. Because fractals occur in different types of sets, various procedures are required to identify and analyze them, with the approach depending upon the kind of data.

Self-Similarity

An object is self-similar when it is composed of smaller copies of itself, and each of those smaller copies is in turn made up of yet smaller copies of the whole, and so on, ad infinitum. The word *similar* carries its geometrical meaning: objects that have the same form but may be different in size. The result is an object composed of a single pattern that repeats itself many times at many different sizes. Moreover, as their size shrinks, the copies of the pattern multiply, so the smaller the size the greater the frequency of the copies. Conceptually, the process is iterative: At each scale, the construction process repeats itself. The proliferation of a single pattern at increasingly smaller sizes within a fractal generates extreme apparent complexity in spite of algorithmic simplicity. Think of a naked tree in winter. The trunk splits into branches that split in turn again and again until reaching the buds on the tips of the smallest twigs. Small branches have approximately the same structure as large branches, except for the difference in size: thus, they are self-similar because the statistics of the small branches mirror those of the large branches. The result is a complex structure of branches that fills a significant portion of three-dimensional space. Many trees are, in fact, fractal structures to within a reasonable approximation. What is the formula for a tree? The idiosyncratic irregularity of a real tree suggests that even a parsimonious description of all its details would be ridiculously long and complicated. Yet a fractal algorithm can capture the overall pattern with great efficiency.

The statistics of branching structures have been studied extensively, particularly in hydrology (Turcotte, 1997). The same kinds of branching processes yield self-similar structures with fractal statistics in many types of phenomena. They seem especially common in biology, for example, the branching in animal bronchial and cardiovascular systems (Bassingthwaighte, Liebovitch, & West, 1994; Turcotte, Pelletier, & Newman, 1998).

Do similar branching processes occur in human affairs? Naturally. Think of hierarchical social structures that have branches, such as corporate structures, governmental bureaucracies, military organizations, taxonomies of occupations, and kinship systems. Are they fractal? In fact, there is growing evidence that many are (e.g., Abbott, 2001, pp. 157–196; Hamilton, Milne, Walker, Burger, & Brown, 2007; White & Johansen, 2005; Zhou, Sornette, Hill, & Dunbar, 2005).

Wichmann (2005) provides a neat example of a fractal branching process from historical linguistics. He examined the sizes of language families as measured by the number of historically related languages comprising each family. Language families form by descent through the successive divergence of languages from a single common ancestor. Divergence occurs when speech communities are separated by social or political processes, such as migration or conflict. When a speech community is divided and communication between the new subgroups is reduced or eliminated, then distinct dialects emerge, which eventually diverge enough to form new, mutually unintelligible languages. These events are usually conceptualized as a branching process by historical linguists, who illustrate them with dendrograms. So, for example, in his data set, the Indo-European language family is composed of 443 distinct languages, whereas Oto-Manguean (a language family in Mexico and Central America) contains 172 languages. Wichmann found that the sizes of language families have a fractal (power law) distribution, and other investigators have detected other fractal properties in the geographic distributions of languages (Gomes, Vasconcelos, Tsang, & Tsang, 1999).

Branching is by no means the only kind of pattern that can be self-similar. Any kind of pattern that is composed of "pieces" of many different sizes but similar forms (either statistically or geometrically) may well be fractal. So, the frequency distributions of sizes associated with many phenomena are fractal. In archaeology, for example, the sizes of artifact fragments are best described by fractal frequency distributions, evidently because the processes of fragmentation are also fractal, as has been theorized by geophysicists (Brown, 2001; Brown, Witschey, & Liebovitch, 2005). Quite a few social phenomena have fractal frequency distributions of sizes, such as the sizes of firms (Axtell, 2001) and cities (De Cola & Lam, 1993, pp. 17–19).

Self-similarity can be exact or statistical. Deterministic systems, which are those that obey specific rules, can display patterns of self-similarity that are perfect and exact. Many such fractals are known in mathematics, but in most empirical data sets, the self-similarity tends to have a random component. Therefore, in real life, most fractal patterns are statistical rather than exact. There are, however, interesting exceptions to this tendency in social science. People can develop intentional constructs that are strictly self-similar. Consider, for example, a military organization with a predetermined and fixed structure: a certain number of companies of a given size per battalion, and so on up through brigades and divisions. Organizationally, it could exhibit exact self-similarity. Nevertheless, although interesting, this kind of perfectly self-similar intentional fractal appears to be rare.

Again, in mathematical theory, the recursion of the fractal element can repeat infinitely so as to create patterns with infinitely fine detail. It is difficult to imagine such a process or pattern occurring in empirical data sets. Real data sets have finite size limits that affect observations, measurements, and calculations. Therefore, we expect to observe fractal behavior only over a certain range of values; the wider the range, the more persuasive, and perhaps interesting, we find the analysis. If the fractal behavior of the data extends over less than a couple of orders of magnitude, then one should question whether a self-similar model is accurate or useful.

Scale Invariance

Self-similarity entails scale invariance, so scale invariance is also diagnostic of fractals. A thing is scale invariant when it has the same characteristics at every scale of observation. As a result, if you zoom in on a fractal object, observing it at ever-increasing magnification, it still looks the same. The relationship to self-similarity is direct and inevitable: Because a self-similar object is composed of copies of itself at every scale, it looks the same at every scale of observation.

Again, as with self-similarity, only perfect mathematical objects look *exactly* the same at *every* scale of observation. Real objects usually exhibit scale invariance statistically within finite size limits. For example, many geological and geographical phenomena exhibit fractal structures, but they can't be scale invariant at sizes greater than the Earth or smaller than the molecules composing the rocks or soils. A wonderfully intuitive example of scale invariance comes from geology. Geologists always put an ordinary object—a rock hammer, a coin, a notebook—into field photographs of rocks, sections, strata, and so forth. Obviously, they do this so the viewer can visualize the scale of the rocks in the photograph. But why can't the viewer tell the scale of the rocks without a visual aid? Because rocks are scale invariant: they look the same regardless of size. You really can't tell the difference between a pebble and boulder without something of known size in the picture.

Power Law Relations

Self-similarity implies a type of relationship called a "power law." For a set to achieve the complexity and irregularity of a fractal, the number of self-similar pieces must be related to their size by a power law. The connection between power laws and fractals is deep and intimate. Power law distributions are scale invariant because the shape of the function is the same at every magnitude. Power law distributions are the only scale-invariant distributions. "The wide applicability of scale invariance provides a rational basis for fractal statistics just as the central limit provides a rational basis for Gaussian statistics" (Turcotte, 1997, p. 39). Power law relations are quite common in human affairs, which helps explain the ubiquity of fractals. For example, both Zipf and Pareto distributions, which have been widely used in the social sciences, are types of power laws. In the following paragraphs, we will describe power laws and their most important properties, but let's start with a simple example.

A power law function commonly takes the simple form $f(x) = Kx^{-c}$, where K is a constant of proportionality and the exponent c characterizes the function or distribution. This should not be confused with an exponential function. A power law function differs from an exponential function because in the latter, the variable of the function occurs in the exponent, thus, $f(x) = Kb^x$. In a power law, the variable, x, is in the base. Obviously, this means that in an exponential relation, it is the exponent that varies, whereas in a power law, the exponent is a constant that characterizes the relationship. The characteristic exponent of a power law captures essential information about that relationship, specifically, how it scales and the processes underlying it. Thus, the scaling exponent is the statistic that characterizes a power law distribution because it is invariant for a particular distribution. The statistics with which we are probably most familiar, such as mean and variance, do not properly characterize a power law distribution because they are not consistent estimators of the parameters of the distribution.

A power law frequency distribution is highly skewed with a long tail (Figures 1.1a and 1.1b). The shape of the tail is important because it tells us about the probability of unusual or extreme events. For example, earthquake sizes have a power law distribution (Turcotte, 1997, pp. 56–66). Unless you're a geologist, you may not care much about the large number of tiny quakes, most of which nobody actually feels, but everyone cares about the giant quakes that can cause enormous loss of life and property. The largest quakes can cause social, economic, and even political disruption. The probability of the large quakes is described by the right tail of the distribution, which is why the tail is important. Earthquakes are not the only natural hazards described by power law distributions. Forest fires and landslides are also power law distributed (Roberts & Turcotte, 1998; Turcotte & Malamud, 2004).

Human disasters commonly have power law distributions, too. The sizes of oil spills, for instance, have power law distributions (Englehardt, 2002). Economic losses from earthquakes and hurricanes have power law distributions, as do human fatalities from hurricanes, floods, tornadoes, and earthquakes (Barton & Nishenko, 2003; Rodkin & Pisarenko, 2008). The intensities of wars, as measured by battle deaths, also obey a power law (Roberts & Turcotte, 1998), and the same is true of the severity of terrorist attacks between 1968 and 2006 (Clauset, Young, & Gleditsch, 2007). Indeed, so many extreme and catastrophic events seem to exhibit power law behavior that an actuary was moved to remark that the "devil sits in the tail of the distribution" (Mikosch, 2005, p. 325).

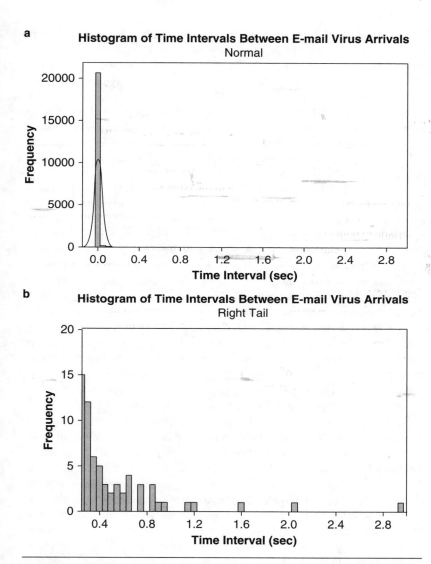

Figure 1.1 *a.* Histogram of a power law frequency distribution, specifically, the time intervals between the arrivals of an e-mail virus. A normal curve is plotted for comparison. The right tail of the distribution is very long, extending across the entire horizontal axis of the graph. *b.* The right tail of the same histogram magnified, showing the region between x = 0.25 and x = 3.0 with a different vertical scale so that the shape of the tail becomes visible.

Price changes in stocks and commodities can also exhibit power law behavior (Stanley et al., 2002). As with earthquakes, you may not be concerned much with small daily movements, but you don't have to be an economist to be intensely interested in the frequency of major market crashes. Recent events have painfully highlighted this problem. When we started writing this chapter a few months ago, the Dow Jones Industrial Average was about 13,000. A few days ago, it closed at 6547, and in the interim, it exhibited stunning volatility. If market fluctuations were normally distributed, such extreme market changes would be ridiculously improbable. In fact, economists' widespread belief in the normal distribution of price changes may have led them to grossly underestimate the risks they were taking, thus leading to the current market crisis (Bouchard, 2008).

Power laws have "heavier" or "fatter" tails than normal *or* exponential distributions. By this we mean that in a power law distribution, the tail encompasses a somewhat larger fraction of the area under the curve than one finds in a normal or exponential distribution. Therefore, extreme values that occur in the tails are not outliers, or anomalies; rather, as we have seen, they are often *the most significant events*. One important reason to identify power law relations is because they can help us to understand the frequency of rare extreme events. Because of their rarity, the frequency of events in the far right tail of the distribution may be difficult to estimate or predict. Their small numbers mean they are strongly subject to random sampling error. But if one knows that the distribution is a power law, one can predict the frequency of extreme events with some confidence. The important principle to remember is that the tails of the power law are often the diagnostic parts of the distribution because they allow one to distinguish between power laws and exponential, log-normal, and other distributions.

By the same token, frequencies at the lower (left) end of the distribution may also be under- or overestimated because of sampling or detection problems. This is the case with small earthquakes, tiny archaeological artifacts, and many other phenomena. Small earthquakes present problems of detection that depend on the density and sensitivity of the seismic detection network (Turcotte, 1997, pp. 39, 57, 60). Tiny archaeological artifacts are difficult to recover. One must pass sediments through progressively smaller screens and/ or use magnification (hand lenses or binocular microscopes) to recover them. Because of their power law frequency distribution, as their size decreases, their frequency dramatically increases. The result is that consistent recovery rapidly becomes extremely time-consuming and uneconomical. Similarly, can you imagine the difficulty, not to mention the bureaucratic costs, of reporting and recording all oil spills smaller than 1 gallon?

A noteworthy and useful characteristic of power law distributions is that they plot as a straight line on a log-log plot (Figure 1.2). That is, if you create a scatterplot in which both the *x*- and *y*-axes have logarithmic scales, the size-frequency relation will plot as a straight line. In fact, this procedure is commonly used as a test of whether a distribution is described by a power law: Plot the frequency distribution on a log-log graph, and if it is linear, then the distribution is approximated by a power law. As an alternative to plotting the data on a graph that has logarithmic scales, one can just as easily take logarithms of both columns of data and plot the logged data on a graph with linear axes. (There is a more general form of power law that includes a periodic component:

$$f(x) = Kx^c g\left(\frac{\text{Log}(x)}{\text{Log}(a)}\right)$$

where K, c, and a are constants and $f(x)$ is a periodic function such that $f(1 + x) = f(x)$ [Liebovitch, 1998, pp. 28–31]. In this case, the power law function on the log-log graph will display a periodic wiggle.) It may be useful to recall in this context that an exponential distribution plots as a straight line on semi-log graph paper, that is, when only one axis of the graph has a logarithmic scale (Figure 1.3). Obviously, this principle can help one to differentiate exponentials from power laws.

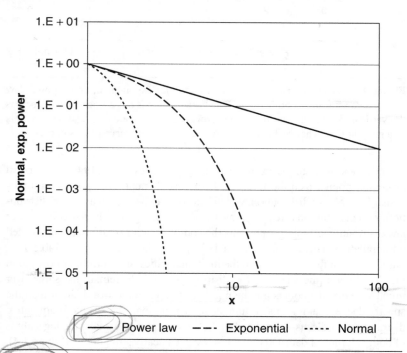

Figure 1.2 A comparison of a normal curve, an exponential, and a power law displayed on a log-log, or double logarithmic, graph, one on which both axes have logarithmic scales. Observe that the power law relation appears as a straight line, whereas the other two are curves.

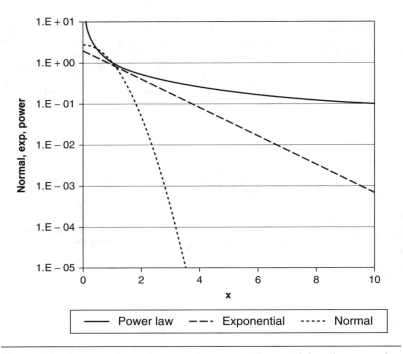

Figure 1.3 A comparison of a normal curve, an exponential, and a power law shown on a semi-logarithmic graph: The horizontal axis has a linear scale, whereas the vertical axis has a logarithmic one. Note that the exponential relationship plots as a straight line, whereas the power law and the normal curve plot as curves.

Let's look at a simple example. George Kingsley Zipf (1949), a Harvard professor, popularized the idea that human settlement sizes possessed a particular kind of size distribution, which has come to be known as a Zipf distribution. Zipf distributions are power laws, and city sizes generally do exhibit power law distributions. Table 1.1 presents the counts of cities of different sizes based on population. These data are from 1970 (De Cola & Lam, 1993, Table 1.5), but it is known that even though the individual cities may shift their position in the distribution, the shape of the distribution tends to remain the same over time. Note that the counts are the number of cities in the world larger than the specified size, making this a cumulative distribution. The issue of cumulative distributions is discussed in detail in the next chapter. If we take the logarithms of both columns of data and plot them, we get a nearly straight line (Figure 1.4). The coefficient of determination for the regression line is > .99, implying that the linear model is a good fit to the data points. The slope of the regression line, 1.14, provides an estimate of the exponent of the power law, although, as we will see in Chapter 2, it is unlikely to be a particularly accurate estimate.

Population Greater Than	Count	Log (Pop)	Log (Count)
4,000,000	24	6.60	1.38
2,000,000	63	6.30	1.80
1,000,000	159	6.00	2.20
500,000	345	5.70	2.54
250,000	726	5.40	2.86
100,000	1615	5.00	3.21

Table 1.1 Relationship between city population and count (frequency), that is, the number of cities larger than the given size in the world, for 1970 (from De Cola & Lam, 1993, Table 1.5, with a correction).

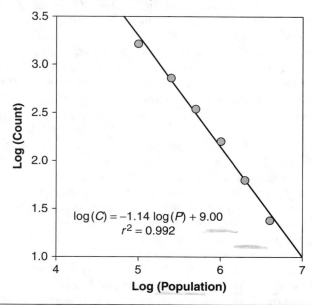

Figure 1.4 Double logarithmic plot of the city sizes and counts in Table 1.1.

We present a second illustration of a double logarithmic plot of a power law distribution in Figure 1.5, which is based on the data in Table 1.2. These are data on the frequency of oil spills of different sizes in navigable U.S. waters in the year 2000. The data are collected by the U.S. Coast Guard under the authority of the Clean Water Act. We downloaded the data from the *2006 Statistical Abstract of the United States* (U.S. Census Bureau, 2006). Individual oil spills can be caused by many different factors (negligence, ignorance, accidents, material failures, storms, ship groundings, etc.). Clearly, the causes of oil spills are complex and interact in multiple ways to create the overall distribution of spill sizes. Note that the size intervals in which spills are recorded are not of constant width, so we need to normalize the values. We do this by dividing each frequency by the width of its bin and by the grand total ($N = 8,354$). It seems likely that the actual volume of many spills would be difficult to measure with precision, and so reporting the data in intervals is reasonable. It is also possible that the format in which data are reported to the government under the various environmental laws, such as the Clean Water Act, permits or requires reporting in size categories rather than in exact numbers of gallons.

The first two columns of data, on the sizes and frequencies of spills, are those provided by the government. The "Width of Bin" lists the range of each size interval. The column labeled "Normalized Frequency" gives the frequency of spills in the row, divided by the bin width divided in turn by the grand total. The "Lower Bound" is simply the lower limit of each bin, and the next column shows the base-10 log of that value. The final column is the base-10 log of the normalized frequency. In Figure 1.5, we have plotted the last two columns. We decided to use the lower bound of the size intervals for two reasons. First, in a case like this, we are most likely interested in knowing about the probability that a spill larger than a certain size will occur, and that idea is represented better by using the left-hand cutoff of the intervals than by using the midpoints of the intervals. Second, using the midpoints of the intervals, which might seem like the natural method of creating a histogram, is really based on the assumption that the data are randomly (normally) distributed within each interval and therefore that the midpoint is probably a good estimate of the mean of those values. In our case, that reasoning is unlikely to hold. As we can see in the table, the data are strongly skewed among the intervals—the narrow intervals in the upper rows have many data points while the wide intervals in the lower rows have very few—and it is likely that the data are similarly distributed within the intervals as well. That is, if this is a power law distribution, and it does seem to be one, then by the principle of scale invariance, we can reasonably assert that within each interval, the data will be skewed as well. So, for example, within the first interval, which runs from 1 to 100 gallons, most of the spills will be small, closer to 1 gallon than to 100 gallons. Thus, the use of the left cutoff provides a better indication of the location of the distribution than the midpoint would.

Size of Spill (gals)	Frequency	Width of Bin	Normalized Frequency	Lower Bound	Log (lower bound)	Log (normalized) frequency
1–100	8,058	99	0.00974310981	1	0	–2.0113024
101–1,000	219	899	0.00002916016	101	2.00432137	–4.53521005
1,001–3,000	37	1,999	0.00000221562	1001	3.00043408	–5.65450554
3,001–5,000	12	1,999	0.00000071858	3001	3.477266	–6.14352602
5,001–10,000	16	4,999	0.00000038313	5001	3.69905685	–6.41665763
10,001–50,000	6	39,999	0.00000001796	10001	4.00004343	–7.74579235
50,001–100,000	4	49,999	0.00000000958	50001	4.69897869	–8.0187958
100,001–1,000,000	2	899,999	0.00000000027	100001	5.00000434	–9.5751065

Table 1.2 Sizes of oil spills in U.S. waters. The raw data (first two columns) are from the Coast Guard, as reported by the U.S. Census Bureau in the *2006 Statistical Abstract of the United States.*

14

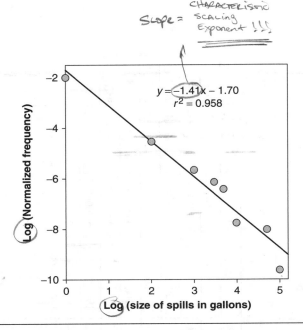

Slope = CHARACTERISTIC SCALING Exponent !!!

$$y = -1.41x - 1.70$$
$$r^2 = 0.958$$

Figure 1.5 Double log plot of the fourth and fifth columns from Table 1.2. Note the strongly linear relation among the data points, which is modeled by the least squares regression line.

In Figure 1.5, the strongly linear relationship among the data points indicates that the distribution is a power law. The coefficient of determination of the ordinary least squares regression line (r^2) is close to 1 (0.96), indicating that the linear fit explains most of the variance in the data. The characteristic scaling exponent is the slope of the line, −1.41. The base of the logarithm used on the graph doesn't matter in estimating the exponent because the slope of the line is, of course, a ratio, and the base of the logarithms doesn't alter the ratio. Base-10 logarithms and natural logarithms (base e) are both commonly used for these purposes, but sometimes other bases are used as well. The choice of the base of the logarithm may be influenced by considerations such as the binning of the data in the histogram, other characteristics of the data, or the nature of the phenomenon being studied. These issues are discussed in more detail in the next chapter.

Another important consideration is the range of values over which the power law relation holds. The greater the range over which we observe fractal behavior, the more likely a fractal process is fundamental to the development of the phenomenon. In Figure 1.5, the power law appears to hold over five logarithmic bins, or five orders of magnitude. A useful rule of thumb is that we would like to see fractal behavior over at least a couple of orders of magnitude in our data set; otherwise, a fractal model may not be a revealing or helpful model of our data.

As we said above, the exponent captures essential information about the pattern in the data and may be thought of as its signature. The exponent quantifies the scaling relationship between the number (frequency) of "pieces" and their sizes (magnitudes). In most fractals, this exponent is a simple function of the fractal dimension, D. In fractals, and especially in empirical, random fractals, the exponent is usually a fraction, not an integer.

At this point, you may be wondering what's "rough" about a power law distribution. The distribution itself looks like a smooth curve, and it can easily be transformed to a straight line. How, then, does it relate to the complexity, irregularity, and roughness that characterize fractals? The distribution is indeed smooth, but we use the distribution to evaluate the rough or irregular characteristics inherent in the phenomenon. We use the distribution to estimate the distribution of the sizes or magnitudes of the elements, pieces, fragments, tendrils, links, or fluctuations comprising the phenomenon. So, for example, when we evaluate the size distribution of cities, we are really investigating how clumpy the population is because cities are clusters of people. When we find that the distribution is a power law, this tells us that the clusters vary greatly in size and have no characteristic magnitude toward which they trend. These qualities are diagnostic of scale-invariant, self-similar patterns that appear very complicated. When we look at a map, this kind of irregularity is evident in the distribution of settlements. Thus, the smooth power law reveals the simplicity of the self-similar, scale-invariant pattern underlying the apparently complex distribution. In a similar way, if you look at a diagram of a fractal network, it appears indecipherably complicated: There are lines running all over. But the study of fractal networks reveals that simple power law relations govern the distribution of links between nodes. Again, a fractal time series looks very irregular. It is in the distribution of the sizes of the fluctuations that we find a power law distribution. In sum, power laws provide much of the algorithmic simplicity underlying the complexity of fractal patterns.

Fractal Dimension: Quantifying Fractal Properties

The concept of "dimension" is elusive and very complex....Different definitions frequently yield different results, and the field abounds in paradoxes. (Mandelbrot, 1967, p. 638, Note 5)

The fourth idea we said we would discuss is the fractal dimension. What is the fractal dimension? In general terms, the fractal dimension is the invariant parameter that characterizes a fractal set. Just as with a "normal" set of data, where we use the mean and variance to describe the location and dispersion of the data, with a fractal set, we use the fractal dimension to describe the distribution of the data.

In traditional Euclidean geometry, dimensions are integers: 0 for a point, 1 for a line, 2 for a plane, and so forth. Modern conceptions of dimension, in contrast, include various ways of measuring dimension that can produce fractions and even negative numbers (Mandelbrot, 2003), and which are, therefore, strictly speaking non-Euclidean. These methods include, for example, the topological dimension, the correlation dimension, the information dimension, the capacity dimension, and others. They are all mathematically related but measure different characteristics of sets. Here we will begin by discussing the simplest version of the fractal dimension. This dimension is described by the following relation:

$$N = \frac{1}{s^D},$$ (1.1)

where N is the number of self-similar "pieces," s is the linear scaling factor ("sizes") of the pieces to the whole, and D is the dimension that characterizes the (invariant) relationship between size and number.

Rearranging the elements of Equation 1.1, we can solve for D:

$$D = -\left(\frac{\log N}{\log s}\right). \quad = -\log\left[N-s\right]$$ (1.2)

Let's explain now how the D in an algebraic equation, like Equation 1.1, can give us a dimension, which is a concept from geometry, not algebra. Let's cut a one-dimensional line into pieces, each of which is a fraction s of the original line. To be clearer, let's give a specific example, and make $s = $ ¼. If we cut our one-dimensional line into pieces that are each ¼ the size of the original line, then we have $N = 4$ little lines. Then, Equation 1.1 tells us that our line has fractal dimension $D = 1$, because $N = 1/(1/4)^1$. If we cut a two-dimensional square area into pieces, the side of each of which is ¼ the size of the original square, then we have $N = 16$ little squares. Equation 1.1 tells us that the area of our square has fractal dimension $D = 2$, because $N = 1/(1/4)^2$. If we cut a three-dimensional cube volume into pieces, the side of each of which is ¼ the size of the original cube, then we have $N = 64$ little cubes. Equation 1.1 tells us that the volume of our cube has fractal dimension $D = 3$, because $N = 1/(1/4)^3$. In fact, no matter the value of s, we still find $N = 1/s^D$ pieces when we cut up one-, two-, and three-dimensional objects. So Equation 1.1 gives us the correct fractal dimension, one that equals our intuitive notion of dimension, for one-dimensional lines, two-dimensional areas, and three-dimensional volumes. Even more interesting, Equation 1.1 can also give us the fractal dimension for other, non-Euclidean objects, the dimensions of which are not the integers that we are so used to thinking must be associated with dimensions.

To clarify the interpretation of this concept of dimension, let's look at some mathematical objects with fractal properties and apply this formula. We'll start with a simple fractal, the middle-third Cantor set. If we take a line and divide it into three equal segments, then the number of pieces, N, is 3 and the linear scaling factor, s, is ⅓. We can then take each segment and divide it into three smaller segments, and so forth. This is a self-similar process, of course, although a simple line is not fractal. The log of 3 divided by the log of ⅓ is −1, and the negative of that is 1. Now, we expect a line to have dimension 1, so all is right with the world. We can see that for this nonfractal object, the calculation of the fractal dimension yields the same value as the Euclidean and topological dimensions.

However, when we divide our line into three segments, what if we throw out the middle segment so that we are left with only two segments? Figure 1.6 illustrates this process through four iterations, starting with the unit line segment. If we repeat this process indefinitely, and we are careful to retain the endpoints of each line segment, we end up with an infinite "dust" of points. Let us calculate the dimension of this object. Our scaling factor of ⅓ is unchanged, but because we are discarding the middle segment, we end up with only two (not three) line segments, each of which is ⅓ the size of the original line segment. Our equation becomes

$$D = -\frac{\log 2}{\log \frac{1}{3}} \approx 0.6309$$

or, alternatively,

$$D = \frac{\log 2}{\log 3}.$$

Because points are supposed to have a Euclidean dimension of 0, this fractional value of 0.6309 may seem strange, but it describes the invariant power law scaling relation of the set. This is actually the dimension of Georg Cantor's famous "Cantor set." Cantor was a late–19th-century mathematician who made fundamental contributions to the foundations of set theory. The Cantor set is the primordial fractal, both historically and mathematically. In fact, there is not just one Cantor set. One can construct a multiplicity of Cantor sets by varying the parameters of the process. For instance, one can divide the line into five segments and retain three, yielding a set with dimension ≈0.6826.

18

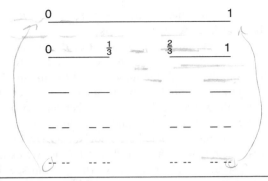

Figure 1.6 The middle-third Cantor set.

The Cantor set illustrates several important properties of fractals. First, it involves a "contraction"—the unit line segment is reduced to ⅓ of its original size. Second, it is produced by an iterative or recursive feedback process. Third, it is self-similar: If you magnify a section, it looks like the whole. Fourth, it exhibits an intricate pattern that is generated by the recursive application of a simple rule. Fifth, the number of pieces $N(x)$ of size x has the power law distribution $N(x) = x^{-\left(\log 2/\log 3\right)}$ because two times as many pieces are created each time the size changes by ⅓ (when $x \rightarrow \frac{1}{3}x$, $N \rightarrow \left(\frac{1}{3}\right)^{-(\log 2/\log 3)}$ $N = 2N$). Sixth, the fractal dimension $D = (\log(2)/\log(3)) \approx .6309$, which is a fraction, not an integer.

For most fractals, their dimension D is a fraction, not an integer. In fact, every set with a noninteger D is a fractal (Mandelbrot, 1983, p. 15), so, by definition, any set with a fractional dimension is a fractal. This is a useful property because it helps us identify fractals. Although some fractals can have an integer dimension (Mandelbrot, 1983), they are rare, particularly with empirical data sets. The fractal dimension measures the roughness and complexity of the set. For example, in a plane, a fractal curve will have a fractal dimension between 1 (the Euclidean dimension for a line) and 2 (the Euclidean dimension for a plane). The more complex the curve, the closer the dimension will be to 2. The theoretical maximum dimension for a fractal curve is 2, when it becomes so complex it fills the plane. Similarly, the larger the dimension of an object embedded in three dimensions, the more a fractal object fills up the space around it. For example, a tree with fractal dimension 2.1 in our common three-dimensional space is a very sparse and spare tree, just a few thin lines very open to the sun and wind. An object with fractal dimension 2.5 is much denser, with many more branches at each joint. An object with fractal dimension 2.9 is thick, with many branches blocking the sun and wind. *Just as the mean describes Gaussian (normal) data, the fractal dimension is the single most important descriptor of fractal objects, processes, and data.* This strange, non-Euclidean conceptualization of dimension is a fundamental characteristic of fractals and is the source of their apparent complexity and naturalness.

To consider a concrete example, we can look back at our oil spill graph (Table 1.2). We can see that the number of self-similar "pieces" in this case is the frequency of the spills, which is plotted logarithmically on the *y*-axis. The sizes are plotted logarithmically on the *x*-axis. The slope of a line is defined as the "rise over the run," which is equivalent to the expression in parentheses on the right-hand side of Equation 1.2. A fractal power law normally will have a negative exponent, because the number of pieces increases as their sizes decrease. The dimension, however, is the negative of the slope, which is a positive number. So, in this case, the fractal dimension of this empirical, random fractal is 1.41. In such cases, the exponent tells us about the distribution of objects or events, but does not always have a clear geometrical interpretation.

The same properties that make the fractal dimension the appropriate parameter to describe a fractal object make the mean and variance unstable measures of the characteristics of the phenomenon. When we sample a population, we generally expect that the sample mean will be an estimate of the population mean. Furthermore, we expect that that estimate will improve as the size of the sample grows. Similarly, we expect the variance of the sample to quantify the spread or dispersion of the data in the population. With some kinds of fractals, this is not the case: The mean and variance are sometimes not stable—they can increase or decrease with sample size without converging on a finite "true" value (Liebovitch, 1998, pp. 74–105; Liebovitch & Scheurle, 2000; Liebovitch & Todorov, 1996). Thus, the sample mean and variance are not consistent estimators of the population parameters in a fractal. Another way to state this is to say that because Equation 1.1 is a highly skewed function, with no resemblance to a normal curve, the mean and variance have little value in describing it. That is precisely why one calculates the fractal dimension instead; it *is* the stable, consistent estimator of the population parameter.

Let's look at another classic example of a fractal to see what these ideas mean in practice. Like the Cantor set, the Sierpiński triangle (or "gasket") is a fractal that has been studied for many years (since 1916) and has played a significant role in mathematics. One starts with an equilateral triangle and then removes another equilateral triangle half the linear size (half the height

20

and width) of the original from the middle (Figure 1.7). This leaves one with three new equilateral triangles touching at their vertices. Then one removes another triangle, half the size of the previous one, from the center of the three remaining triangles. This rule is iterated. Another equivalent way of thinking of this process is as a transformation: Take the original triangle, scale it by half, copy it twice, and translate (move or shift) those copies to the appropriate corners. Then, do that again to the new figure, and iterate that rule. As Figure 1.7 illustrates, this process produces a highly complex pattern composed of triangles of every scale whose frequency is related to their size. The fractal dimension of the Sierpiński triangle can be calculated from Equation 1.2. The scaling factor is ½ and the number of "pieces" is 3 at each iteration. Thus,

$$D = -\frac{\log 3}{\log \frac{1}{2}} = 1.585 \ .$$ (1.2)

In sum, fractals are sets that exhibit self-similarity and scale invariance. These properties logically entail the existence of power law structures in the data. The fractal dimension, in turn, is the statistic that allows us to infer the invariant parameter of a fractal data set. The mean and variance of fractal data sets are, at best, poor and unstable estimators of the parameters of fractals. The fractal dimension quantifies how the characteristics of a fractal scale in number, magnitude, space, and/or time. All sets with a fractional fractal dimension are fractals.

The Formal Definition of Fractals

Mandelbrot (1983, pp. 361–362) long resisted supplying a strict and formal definition of fractals because no single definition encompassed all the patterns that seemed intuitively to exhibit the essential qualities of fractals. The

Figure 1.7 The first four iterations of the Sierpiński Triangle. The black parts form the fractal.

diversity of fractal phenomena makes a single, simple definition elusive, but Mandelbrot's formal definition is a key to understanding them.

Mandelbrot (1983) wrote that a *"fractal is by definition a set for which the Hausdorff Besicovitch dimension strictly exceeds the topological dimension. Every set with a noninteger D is a fractal"* (p. 15, emphasis in original). To apply this definition, we need to understand the topological and Hausdorff Besicovitch dimensions. We describe those dimensions below, but if you find the following explanations challenging, we encourage you to skip to the next section because you don't need to understand some of the more difficult bits just yet.

Topological Dimension

The topological dimension "describes how the points that make up an object are connected" (Liebovitch, 1998, p. 60). The most important facts to remember about the topological dimension are as follows:

1. It is always an integer; and

2. The topological dimensions of lines, planes, and volumes are the same as their Euclidean and Cartesian dimensions, that is, 1, 2, and 3, respectively.

The topological dimension can be defined in several different ways, which are not so easy to describe. One, perhaps more intuitive, definition depends on realizing that the boundaries of an object always are one dimension smaller than the object itself. For example, the three-dimensional room around you now is confined by the borders of the two-dimensional areas that make up the floor, ceiling, and walls around you. If you look at any one of those walls, its borders are one-dimensional lines. And if you look along any of those lines, a position in it is bordered by two zero-dimensional points. The topological dimension can be defined as the number of times we need to take the borders of the borders (and keep repeating that process) until we get to those last two zero-dimesional points. Because this takes three steps for the room that you are in (2-D walls, 1-D lines, 0-D points), the topological dimension of your room is 3 (Liebovitch, 1998, pp. 60–61).

Hausdorff Besicovitch Dimension

The Hausdorff dimension was developed by German mathematician Felix Hausdorff around 1918 in order to resolve problems and inconsistencies in earlier definitions of dimension. The concept was significantly expanded and elaborated by Besicovitch and his students in subsequent decades, hence the double-barreled name used by Mandelbrot. In fact, it is more commonly called just the Hausdorff dimension.

Let's sneak up on the Hausdorff dimension by first describing Kolmogorov's capacity dimension. To calculate the capacity of a set, we cover it with spheres of radius r. Then we determine the minimum number of spheres $N(r)$ of that size required to cover all and only the set or object. Next, we shrink the size of the spheres and recalculate the minimum number needed to cover the set. We repeat the procedure a number of times. The number of spheres required will be inversely proportional to their size. Consider a line segment of length 1. It can be covered by one sphere of diameter 1, two spheres of diameter ½, three spheres of diameter ⅓, and $N(r)$ spheres of diameter $1/r$. The capacity dimension will be the relationship between the size of the spheres and the number required for the covering in the limit as the diameter of the spheres approaches 0 (Liebovitch, 1998, pp. 50–51):

$$D_c = \lim_{r \to 0} \frac{\log N(r)}{\log \frac{1}{r}},$$

(1.3)

where $N(r)$ is the number of spheres of size r. For our trusty unit line segment, Equation 1.3 equals 1. Equation 1.3 is a closely related generalization of Equations 1.1 and 1.2.

The Hausdorff dimension is in turn similar to the capacity dimension. The rigorous determination of the Hausdorff dimension is fairly complicated, and here we need only present a simplified description of it. As with the capacity dimension, we produce a covering of the set or object we want to measure, but instead of covering it with spheres, we cover it with sets that can vary in size. We then minimize a function that is the sum of the diameters r of the covering sets raised to a power s. This procedure is repeated using sets smaller than a given size. The function will approach a limit as the maximum size of the covering sets nears 0. The limit will vary depending on the value of the exponent s. When s is below a certain critical value, the function will grow very large and its limit will approach infinity. When s is above the critical value, the function will shrink rapidly and its limit will approach 0. The critical point marking the change in the behavior of the limit of the function, where it jumps from ∞ to 0, is the Hausdorff dimension (Liebovitch, 1998, pp. 52–53). Fuller discussions of the Hausdorff dimension can be found in many books on fractals, as well as in general books on topology and measure (Falconer, 2003, pp. 27–37; Peitgen, Jürgens, & Saupe, 1992, pp. 215–219). Note that the capacity dimension is a special case of the Hausdorff dimension in which the diameters of the covering spheres are constrained to be the same.

The Hausdorff dimension has a number of useful properties. It normally quantifies dimension in a way that reflects our intuitive understanding of the concept. Conveniently, the Hausdorff dimensions of simple Euclidean objects such as lines, squares, circles, and cubes correspond to their respective topological dimensions. Therefore, those objects are not fractal because their Hausdorff dimensions do not exceed their topological dimensions. For more complicated sets such as fractals, the Hausdorff dimension does exceed the corresponding topological dimension, usually by a fraction, but sometimes by an integer.

The Hausdorff dimension provides the most rigorous mathematical characterization of fractal dimension, but in practice it is rarely used. Although it is conceptually elegant, it is difficult to calculate in many concrete cases. Therefore, most scientists use a variety of other approaches to estimate the fractal dimensions of empirical phenomena. These approaches include the self-similarity, capacity, and box-counting dimensions. These are used much more commonly than the Hausdorff dimension, and we will discuss them in the next two chapters.

Discussion

So far, we have seen that fractals are rough and irregular sets characterized by self-similarity; scale invariance; and, almost always, fractional dimension. We have described, albeit briefly, several ways of evaluating dimension, including the topological, capacity, self-similarity, and Hausdorff dimensions. We have examined several examples of complicated sets that yield fractional dimensions. We reviewed the formal definition of fractals as sets for which the Hausdorff dimension strictly exceeds the topological dimension. We have noted that traditional parametric statistics are not appropriate for measuring fractal objects. Does this imply that fractal analysis is relevant to your work? Should you plow on? In deciding, consider the following points.

First, fractal patterns abound in cultural behavior and social relations. In fact, fractals provide a significant organizing principle of human life. This is a bold assertion, but we will support it throughout the book by providing a variety of examples. We do not assert, however, that *everything* is fractal. Many sociocultural phenomena are clearly not fractal. Therefore, each case should be evaluated carefully on its merits, and we should reject unsupported, broad-brush claims for fractal patterns. We do not know, for example, whether fractal analysis represents a Kuhnian paradigm shift. That is a matter for future historians and sociologists of science to debate.

By the same token, claims for the fractality of social or cultural patterns should not be judged by unrealistic criteria that would not be applied to other kinds of analyses. Such claims are categorically *not* extraordinary and

→ my RESEARCH

therefore do not require extraordinary proofs. Finding a fractal pattern in social data should not be considered any more remarkable than finding a normal distribution, although even the use of the term *normal* biases us against accepting other kinds of patterns.

For the moment, let's note that fractals are ubiquitous in the physical and natural sciences. At small scales, bacteria form fractal colonies, and at the largest scale, some think that the distribution of mass in the cosmos is fractal. We are fractal—our lungs, nerves, and veins are fractals, and even the beating of our hearts can be a fractal time series. Searches of citation databases and library catalogs reveal thousands of articles and hundreds of books on fractals.

Fractals occur in diverse natural phenomena with such frequency that many scientists believe they possess a quality called *universality*: the tendency for similar outcomes to emerge from dynamical systems or models despite differences in their specifications. In turn, universality implies that the details of models are often insignificant because overall, the systems exhibit characteristic behavior. In other words, if similar patterns keep popping up under widely different circumstances, then clearly some generic principle must be operating, one that does not depend on the specifics of each case. The appearance of fractals in many different social science phenomena also implies considerable universality, and therefore that common kinds of mechanisms or processes are at work. Fractals are thus a unifying theme among the human sciences, and, wishing to emphasize this, we have chosen the examples in this volume from a broad range of disciplines.

Because fractals are common, we should understand them so that we can identify and analyze them correctly. We are often not prepared to recognize fractal patterns because we have not been primed by our training and education to see them in our data, despite the fact that they often mimic patterns in nature. We are generally trained to see normal bell curves and linear patterns; in fact, we are taught to create bell curves or linear relations through data transformations or smoothing. But fractals are neither normal nor linear. They are strongly nonlinear and non-normal, yet this does not mean that they are either badly behaved or random. They are strongly patterned in their own right, but in a strange way, a pattern to which we are not generally accustomed. Therefore, seeing fractals requires training, practice, and imagination, but when one begins to discover such patterns in one's own research, the effort is rewarded. → YES!!! ✱✱✱

II. Second, fractals are complex patterns that can be described by simple algorithms. Therefore, they are intellectually interesting and appealing because they provide parsimonious descriptions of complex phenomena. Although parsimony does not necessarily trump all other characteristics of scientific explanation, it is highly desirable, and parsimony may help you decide among alternative models. If you have ever struggled to model your data with increasingly complex models that yield only incrementally better coefficients of determination, your data may be a candidate for fractal analysis. "These [fractal] structures present a maddening challenge to our methods, because they contain information that cannot be captured in linear form" (Abbott, 2001, p. 165). Furthermore, as Abbott also points out, fractal properties and processes in society are undertheorized, which means that this is a fertile field for exploration.

In particular, data sets and processes that have been modeled using other probability distributions, such as the normal, the Poisson, the exponential, or the lognormal, sometimes turn out to match to a power law (fractal) distribution much better. For example, in studying offense rates of criminals, Spelman (1994) found that many were fit well by Pareto distributions, a form of power law. Nevertheless, he observed that,

> unfortunately, the mean of a Pareto distribution is infinite when the shape parameter is less than 1. The shape parameters of the fitted Paretos are consistently less than 1.... So, although the Pareto form fits the empirical data well, it would be very awkward to use. (pp. 118–119)

So he chose instead, "more or less by default," to use the lognormal distribution in his model. In such cases, the appropriate fractal analysis may provide greater insight by revealing the true parameters of a data set. These parameters, in turn, help infer and measure vital aspects of the underlying phenomena that create the data sets.

Third, fractals are intimately related to nonlinear dynamical processes such as cellular automata, self-organized critical systems, and iterated function systems. Many social scientists are interested in investigating these types of dynamical systems in society, and fractal analysis provides one method for approaching such investigations.

Finally, fractal analysis remains fresh, exciting, and fun. Regression with dummy variables may be gratifying to the specialist, but "fun" is not a word that comes to mind when describing it. Yet many practitioners describe deep enjoyment in fractal analysis because of its elegance, artistry, and parsimony. Moreover, many low-hanging fractal fruit are waiting to be picked in many fields, but particularly in the social sciences, where fractal analysis has not yet become ubiquitous. This is why fractal analysis is one of the fastest growing and most exciting fields in science. Because it is so fun, interesting, and accessible, if we were graduate students now, we would be looking at fractal analysis to develop new research projects.

CHAPTER 2. FRACTAL ANALYSIS OF FREQUENCY DISTRIBUTIONS

In his seminal classic on fractals, *The Fractal Geometry of Nature* (1983), Mandelbrot wrote,

> If monographs or textbooks on fractals come to be written, the discussion of random geometric shapes, which is mathematically delicate, will come after the less difficult topic of random functions, and these books will begin with random variables. On the other hand, this Essay plunges straight into the most complicated topic, because it the most interesting one, and gives play to geometric intuition. (p. 341)

Many monographs and textbooks on fractals have now been written, but contrary to Mandelbrot's prediction, they usually start, as he did, with the geometric constructions, probably because they are more visually intuitive. We, on the other hand, will do as he said, not as he did, and begin with random variables, functions, and distributions. We made this choice because we believe that the statistical training and practice of social scientists emphasize variables, functions, and distributions more strongly than geometrical constructions. Let's begin, then, with power laws.

Power Laws

We now describe how to analyze data to determine if they are fractal. In the previous chapter, we explained that the frequency distribution of fractal data is typically a power law. The mathematical construct that we want to determine from the data set is called the *probability density function* (PDF). The probability that values of x in the data are found in the range (x, $x + dx$), where dx is very small, is given by the integral of the PDF from x to $x + dx$. For fractal data, PDF(x) is proportional to x^{-c}. (More complex fractal scalings are also possible [see Liebovitch, 1998, pp. 28–31], but we will discuss here only this most basic fractal scaling.) On a logarithmic-logarithmic plot—that is, log PDF(x) versus log(x)—this relationship is a straight line with slope equal to ($-c$). Thus, the bottom line is simple: We construct a logarithmic-logarithmic plot of the PDF from the data and see if it looks like a straight line. If it's close to a straight line, then the data are fractal; if not, the data have a different form.

The PDF is often estimated by forming the frequency histogram of how often data values are found within consecutive bins of constant width. This

26

method has become so commonplace that it may be surprising to learn that it is not always the best or most accurate way to determine the PDF. In fact, determining the PDF, and even further determining the best fit of different mathematical forms to the PDF, can be quite challenging. In the following examples, we describe different methods to determine the PDF and illustrate their respective merits and weaknesses.

We begin by describing three different methods to estimate the PDF. In order to give the reader a realistic understanding of how these methods actually work in practice, we will show the results of using them to analyze two different types of data sets. First, we will see how they fare when used to analyze test data sets with a large amount of data of a known form that were generated with known parameters. Then, we will show what happens when they are used to analyze a real-world data set with a much smaller amount of data whose form may not be so pure.

Three Methods: Histogram PDF,
Multiscale PDF, and Cumulative Distribution

Histogram PDF ——> Not Reliable !!

A common way to determine the form of the PDF is to evaluate the frequency distribution by finding the number of values of the data, $n(x)$, that lie in the "bin" from x to $x + dx$ (Spiegel & Stephens, 2008). The centers of these bins can then be used to provide points on the PDF, namely,

$$PDF(x) = n(x)/(dx\ N), \tag{2.1}$$

where dx is the size of each bin and N is the total number of values in the data set. A Matlab program to compute the histogram PDF from data values stored in an ASCII text file, separated by returns, is available on Professor Liebovitch's Web site. In principle, this seems pretty simple. In practice, choosing the size of the bins to best represent the PDF can actually be quite challenging. For many data sets, there are fewer large values than small values in the data. This is especially true for fractal data sets, whose values span many orders of size, with an ever smaller amount of data at ever larger data values. As shown in Figure 2.1, if we choose narrow bins to capture the fine details of the PDF for the small values, then many bins at large values will likely contain 0 or 1 data values, making it difficult to determine the PDF at those large values. On the other hand, if we choose wide bins, so that there are many values in each bin to better capture the shape of the PDF at large values, then we group all the small values together, providing only a very coarse picture of the PDF at the small values.

28

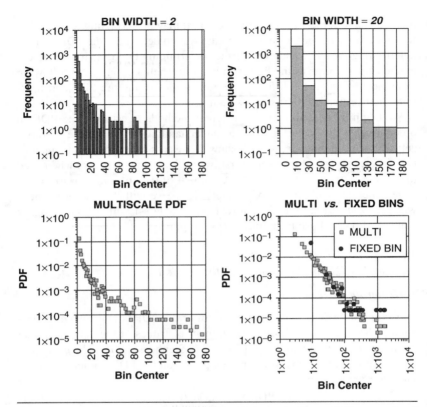

Figure 2.1 Histograms of a data set of 2,048 values of x from a PDF proportional to x^2, formed from bins of width 2 (top left) or 20 (top right). The narrow bins of width 2 capture the fine details at the small values but not at the large values. The wide bins of width 20 give only a coarse view at the small values but provide a good picture at the large values. The multiscale PDF (bottom left), which combines both narrow and wide bins, provides good details at both the small and large values. As shown together (bottom right), the multiscale PDF (gray) defines the distribution more clearly than the histogram PDF (black squares).

Multiscale PDF

Histograms with narrow bins are best at determining the PDF at small values of the data, and histograms with wide bins are best at determining the PDF at large values of the data. Therefore, to get the best of both worlds, we can combine histograms of both narrow and wide bins. The only problem is that we can't combine histograms—but we can combine PDFs. So, we first transform each histogram into a PDF, and then combine the PDFs. This method of determining the PDF is more reliable and more accurate, and it typically covers a much larger range of values than the PDF determined from one histogram with a fixed bin size, as shown in Figures 2.1 and 2.2.

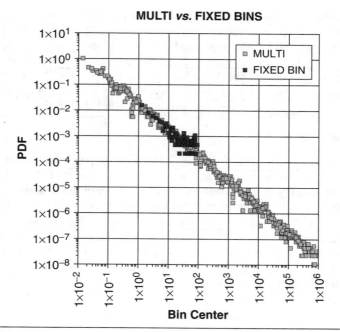

Figure 2.2 In this data set of 5,120 values of x from a PDF proportional to x^{-1}, the multiscale PDF (gray), which combines small and large bins, extends over a very much larger range than the histogram PDF (black) of a fixed bin width.

In the procedure developed by Liebovitch and his colleagues (Liebovitch, Scheurle, Rusek, & Zochowski, 2001; Liebovitch et al., 1999), a series of histograms is formed from the data, each with bins twice the width of the previous one. The first bin in each histogram is dropped because it clumps together a large range of values (from zero to the width of the bin) in this one bin. The bins past the 20th bin are also dropped because the number of large values contained in these bins is typically small. (There is nothing particularly magical about choosing exactly 19 bins. We use this number because our experience, from many different data sets, is that the number of data values is typically quite low when the number of bins exceeds 20.) Next, each remaining histogram is transformed into a PDF by using Equation (2.1), and the PDF(x) from each remaining bin from each histogram is plotted versus the center of each bin. A Matlab program to compute this multiscale PDF is available on Professor Liebovitch's Web site. It works very well with long-tailed distributions, that is, when the average number of values decreases with the size of the values. Because it is designed to pick out the structure of that tail, it does not work well when there is a strong local maximum in the data values. Because more points in the PDF are generated where the most data occur in the overlapping

Such as the case here!!

histograms, this provides an intrinsic weighting to more accurately determine the parameters of a functional form used to fit the PDF. Please note that the only assumption this method makes about the data is that the average number of values decreases with the size of the values. It does not make any other assumptions about the form of the PDF. Hence, this method is quite good at detecting when the PDF has a variety of different forms, such as exponential $\exp(-kx)$, stretched exponential $\exp(-kx^c)$, or power law x^{-c}.

Rank CDF

The *cumulative distribution function* (CDF) is the probability that a value larger than x is found in the data. (In many applications, the CDF is defined as the probability that a value less than x is found in the data, that is, the left-tail probability. However, following Mandelbrot [1977, p. 104], we define the CDF as the right-tail probability that a value is greater than x, because it is typically the largest values of fractal data that form the power law tails of these distributions.) The CDF is the integral of the PDF from x to the largest value in the data set, and, complementarily, the PDF is the derivative of the CDF, namely,

$$PDF(x) = -d\,[CDF(x)] \,/\, dx. \tag{2.2}$$

Thus, we can also determine the PDF from the derivative of the CDF. Even better (as we shall see later), there is sometimes a simple relationship between the mathematical forms of the PDF and CDF. In that case, we don't even have to take this derivative; instead, we can work directly with the CDF determined from the data. Using the CDF may not seem like much of an advantage because we are faced with all the same binning challenges in determining the CDF as the PDF. However, there is a neat trick to determining the CDF without binning the data! This method doesn't seem to have a single well-accepted name, and different authors have referred to it as the "empirical cumulative distribution function" (Rice, 1988), the cumulative distribution "calculated by ranking by size" (Edwards et al., 2007), or the "cumulative rank-frequency distribution" (Clauset, Shalizi, & Newman, 2007). The trick is to note that if we sort the values of data in increasing size from the smallest data value of rank = 1 to the largest data value of rank = N, then the probability $CDF(x)$ that the data have a value greater than x is equal to $[N - rank(x)]/N$. Thus, if we plot these values of $CDF(x)$ on the vertical axis versus their corresponding value x on the horizontal axis, we have the CDF without any bins. This method is illustrated for an example of five values of x in Figure 2.3. A Matlab program to compute this rank CDF is available on Professor Liebovitch's Web site (although it is just as simple to use a spreadsheet program, like Microsoft Excel, to compute the CDF this way).

Other Methods

In this monograph, we concentrate on the most basic methods to understand the most salient features of the data. Therefore, we have focused on determining the PDF because it provides the social scientist a good quantitative characterization of the data, which may then lead to insights on the mechanisms that generated the data. When a priori knowledge, or intuition, or just plain bias, suggests that the data should have a particular (or peculiar) mathematical form, other methods can be used to sensitively judge the relative goodness of fit of that form, or competing forms. Typically, these methods, such as maximum likelihood estimation or the Akaike information criterion, analyze the data values themselves instead of first forming the PDF. Excellent theoretical and practical descriptions of these methods may be found in Rice (1988), Edwards et al. (2007), or Clauset,

Data	Sorted	Rank	Cumulative Probability $> X = (5-Rank)/5$
137	38	1	0.8
56	50	2	0.6
50	66	3	0.4
88	88	4	0.2
38	137	5	0.0

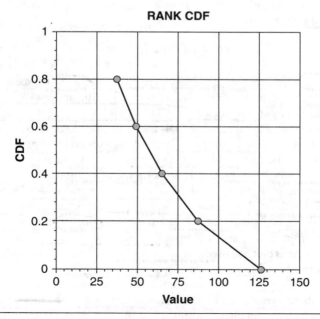

Figure 2.3 The rank CDF is computed from the sorted values of *x* without the need to first form the data into bins.

Shalizi, and Newman (2007). Our own (admitted) prejudice suggests two warnings before the reader is seduced by the (often beautifully rigorous) technical details of these approaches. First, there are more unresolved questions than one might realize about the mathematical issues concerning how to interpret the goodness of fit. (For example, Liebovitch and Tóth, 1990, note that some expansions representing the same mathematical function converge rapidly and others slowly, so insisting that each additional parameter improve the goodness of fit in a statistically significant way does not reliably determine the number of relevant parameters.) Our second warning is to always remember that the goal of fitting functional forms to data is to learn something either about the data themselves or about the mechanisms that generated them, rather than simply to achieve the best goodness of fit.

Examples With Lots of Data

A Power Law PDF With Exponent Equal to 2

We now illustrate the results of using the histogram PDF, multiscale PDF, and rank CDF methods to analyze both large and small data sets of different types. First, we consider a fractal data set of 2,048 values of x that were generated with a PDF proportional to x^{-2}. This was done by 2,048 iterations, each choosing a random value r from a uniform distribution over $(0,1)$ and then setting each value of

$$x = \{(1 - r)[b^{(1 - c)}] + r[a^{(1 - c)}]\}^{[1/(1 - c)]}, \qquad (2.3)$$

where $c = 2$ is the exponent of the power law PDF and a and b are the lower and upper limits, respectively, of the PDF, in this case $a = 10^{-5}$ and $b = 10^{5}$. (Equation [2.3] is derived by solving for x in $r = \mathrm{CDF}(x) = [b^{1-c} - x^{1-c}]/[b^{1-c} - a^{1-c}]$.)

The histogram PDF, multiscale PDF, and rank CDF determined from these data are shown in Figure 2.4. Notice that all the axes on these plots are logarithmic. When the data lie on a straight line on such plots, it means that the PDF has a power law form that is proportional to x^{-c}. In the histogram PDF and multiscale PDF plots, the slope of the line is equal to $-c$. The CDF is the integral of the PDF. Because the integral of x^{-c} is proportional to x^{-c+1}, the slope of the line in the rank CDF plot is equal to $-c + 1$. So, to obtain the exponent of the power law PDF from the CDF, subtract 1 from the slope of the CDF. In each plot, we used a least squares fit of the logarithmically transformed variables to determine the slope of the straight line.

How do the results of the histogram PDF, multiscale PDF, and rank CDF on this data compare? The PDF is not well captured by the histogram PDF method, which cannot choose a single bin size that does a good job at both small and large values of the data. This is reflected in the bins at large values

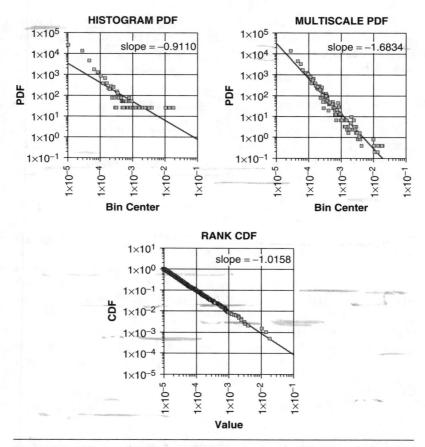

Figure 2.4 Histogram PDF, multiscale PDF, and rank CDF from a data set of 2,048 values of x from a PDF proportional to x^{-2}.

that have zero, one, or two data values in them that form the broken horizontal lines at the lower right of the plot. As a result, the accuracy of the exponent $c = 0.91$ is poor compared to its true value of 2.00. The multiscale PDF is much better in that it extends the range of the PDF considerably, clearly defines it as a straight line, and produces a more accurate $c = 1.68$. The rank CDF is best at defining the shape of the PDF, except at the largest values of the data, and it is the most accurate in determining the slope, which at 1.02 indicates that the exponent of the PDF is 2.02. The conclusion, from this data set, is to rank these methods as follows: rank CDF > multiscale PDF > histogram PDF. If the rank CDF is best, why not use it all the time? We will see in the following examples that a fractal PDF is always a power law and easy to identify as a straight line on a PDF plot of log PDF(x) versus log (x), whereas the CDF is not always a straight line on a plot of log CDF(x) versus log (x), and so the true functional form of the data is harder to identify on a rank CDF plot.

A Power Law PDF With Exponent Equal to 1

You might notice that Equation (2.3) gets nasty when $c = 1$, as the exponents with $1 - c$ and $1/(1 - c)$ become 0 and 1/0. Other methods must be used to generate test data from a PDF proportional to x^{-1}. One simple method is to use the Weibull distribution

$$p(x) = qsx^{s-1}\exp(-qx^s) \qquad (2.4)$$

with parameters $s = 0.05$ and $q = 20$, which provides a good approximation to x^{-1}. The multiscale PDF and rank CDF analysis of such a data set of 2,048 values is shown in Figure 2.5. The PDF is a straight line as expected and has a slope of -0.99, which is close to the expected -1.00. But the CDF is not a straight line at all! The integral of x^{-1} is not a power law, but rather $\ln(x)$. Because the CDF is the integral of the PDF, when the PDF is proportional to x^{-1}, then the CDF is no longer a power law.

The take-home lesson here is that a fractal PDF proportional to x^{-c} is always a straight line on a PDF plot of log PDF(x) versus log (x), but a fractal PDF may not necessarily be a straight line on a CDF plot of log CDF(x) versus log (x). We've seen here that this is the case when $c = 1$. As illustrated by the following example, this is also the case when $c < 1$.

A Power Law PDF With Exponent Equal to 0.5

We also used Equation (2.3) to generate a data set of 2,048 values of x with a fractal PDF of $c = 0.5$. The multiscale PDF and rank CDF analyses of this data set are also shown in Figure 2.5. Once again, the PDF is a straight line on a plot of log PDF(x) versus log (x), but the CDF is not a straight line on a plot of log CDF(x) versus log (x). This will also be the case whenever $c < 1$.

Fractal Data Have a Power Law PDF, But May Not Have a Power Law CDF

The point of the preceding examples with fractal PDFs proportional to x^{-c}, with $c = 2$, 1, or 0.5, is that this type of data is always a power law, which is a straight line on the plot of log PDF(x) versus log (x). This makes it easy to identify this fractal form in the data. However, for the same data, the CDF is

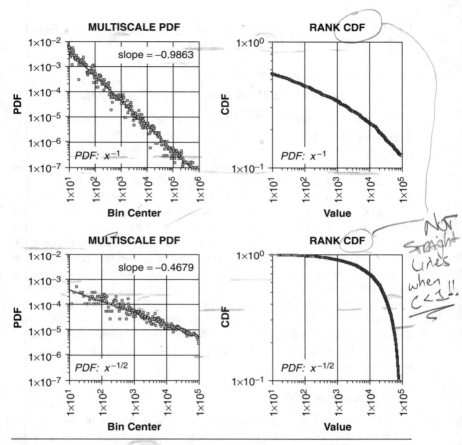

Figure 2.5 Multiscale PDF and rank CDF of a data set of 2,048 values of x from a PDF proportional to x^{-1} (top) and $x^{-(1/2)}$ (bottom). The PDFs are straight lines on these plots, whereas the CDFs are curved.

not always a power law, which would be a straight line on a plot of log CDF(x) versus log x. It is only a power law when the exponent $c > 1$. This makes it more valuable to use the PDF plots, and most especially the multiscale PDF plots, because they make it easier to identify the form of the data.

The formal mathematical details are that if c is not equal to 1 and the PDF extends over the interval (a,b), with PDF$(x) = [(1 - c)/(b^{1-c} - a^{1-c})] x^{-c}$, then the CDF $= (b^{1-c} - x^{1-c})/(b^{1-c} - a^{1-c})$. To normalize the PDF for $c > 1$ requires that a remain small and finite, as b approaches infinity. In that case, the CDF approaches $(x^{1-c})/(a^{1-c})$, which is a power law. However, to normalize the PDF for $c < 1$ requires that a can approach zero, and b remains large and finite. In that case, the CDF approaches $1 - (x^{1-c})/(b^{1-c})$, which is not a power law. For $c = 1$, the PDF = $(1/x)/[\ln(b/a)]$ and the CDF = $\ln(b)/\ln(b/a) - \ln(x)/\ln(b/a)$. Thus, for a fractal PDF proportional to x^{-c} the CDF is a power law only if $c > 1$.

Examples With a Small Amount of Data

Areas of Archaeological Sites in Palm Beach County, Florida

The previous examples all used an extensive amount of data—namely, 2,048 values—taken from a perfectly fractal distribution with known parameters. Many data sets may have a much smaller amount of data whose form may be fractal, or approximately fractal, or not even fractal at all. We now show what to expect when these methods are used to analyze such data. Figure 2.6 shows the results of using the multiscale PDF and rank CDF to determine the distribution of the areas of 189 archaeological sites in Palm Beach County, Florida. Once again, the histogram PDF itself provides only a pale picture of the true shape of the PDF. The multiscale PDF shows some significant scatter around a clear power law relationship, a straight line on this logarithmic-logarithmic plot, with slope −1.18. The rank CDF further shows that this relationship is perhaps more curved than a simple straight line, with overall slope of −0.49, which corresponds to a slope of −1.49 on the PDF. The conclusion here is that the multiscale PDF informs us that, indeed, a fractal power law scaling is a good overall approximation to the data, whereas the rank CDF informs us that the relationship is not an exact power law.

An Exponential PDF

In the example above, we found that the distribution of the sizes of archaeological sites in Palm Beach County is approximately fit by a fractal power law PDF distribution. Can those data perhaps be better fit by a different functional form? A fractal distribution is scale independent and extends in a self-similar way over many scales. A good foil to that distribution is to compare it to a distribution with a single fixed scale. The fractal power law distribution has the form that the PDF is proportional to x^{-c}. An exponential distribution, with fixed scale $(1/k)$, has the form that the PDF is proportional to e^{-kx}. An exponential data set, with the same number of data values as that in the Palm Beach County site data, with the value of $k = 0.0182$ determined from the best fit of an exponential distribution to that data, was generated by choosing 189 values of r equally spaced from $1/189$ to 1, and then setting each value of

$$x = -(1/k)\ln(r). \tag{2.5}$$

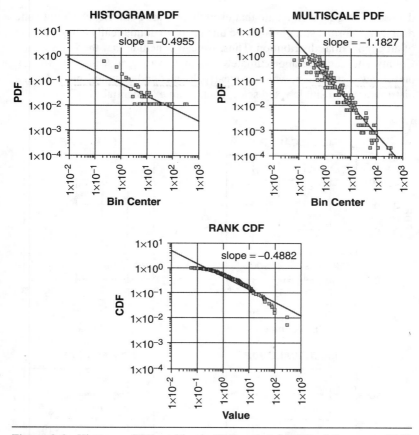

Figure 2.6 Histogram PDF, multiscale PDF, and rank CDF of the areas of 189 archaeological sites in Palm Beach County, Florida.

Equation (2.5) is derived by solving for x in $r = \text{CDF}(x) = e^{-kx}$. Unlike the previous cases, where we chose a random value r from a uniform distribution over $(0,1)$, because the data set here is small, it is better to use equally spaced values of r to provide an exponential distribution without the additional error produced by the randomness in r. The results of using the multiscale PDF and rank CDF to analyze these data sets are shown in Figure 2.7. The top two graphs present the analyses of the archaeological data, and the bottom two graphs present the same analyses for the fitted exponential

38

data set. The graphs illustrate that even though the Palm Beach County site data do not form a straight line, the line is not so dramatically curved as that of an exponential distribution. Thus, we can conclude that the Palm Beach County site data are approximately, but not exactly, a scale-free, fractal, power law distribution, which is clearly distinct and distinguishable from a distribution of a single fixed scale, such as an exponential distribution.

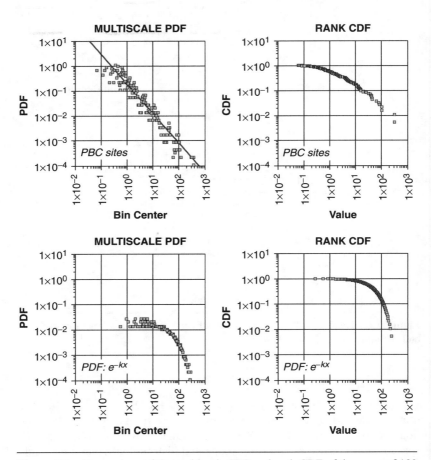

Figure 2.7 Comparison of the multiscale PDF and rank CDF of the areas of 189 archaeological sites in Palm Beach County, Florida (top) with a data set of 189 values of x from a PDF proportional to e^{-kx}, where k was determined from the best fit to the archaeological data (bottom). The PDF and CDF of the PBC archaeological data are more like a power law (straight line) than the highly curved e^{-kx} PDF and CDF.

Summary

The probability that values x in the data are found in the range $(x, x + dx)$ is given by the integral of the probability density function (PDF) from x to $x + dx$. The cumulative distribution function (CDF) is the probability that values of x or greater are found in the data set. The PDF and CDF are the best basic ways to identify fractal data sets. The shape of PDF and CDF plots can reveal if the data set is fractal, approximately fractal, or not fractal. A fractal data set typically has PDF proportional to x^{-c} and so appears as a straight line on a plot of log PDF(x) versus log (x) (although more complex fractal scalings are also possible). A fractal data set, with $c > 1$, has CDF(x) proportional to x^{1-c} and so appears as a straight line on a plot of log CDF(x) versus log (x). There are different ways to determine the PDF and CDF from experimental data. The multiscale PDF and rank CDF are the best methods to determine the PDF and CDF. We prefer the use of the multiscale PDF. This method can be less accurate in determining the value of the parameter c than the rank CDF. However, PDF methods are less ambiguous because fractal data sets are always power laws on these plots, whereas, depending on the value of c, they may or may not be power laws on CDF plots. There are other methods, without forming the PDF or CDF, that quantify the goodness of fit to the data values themselves. These methods can be useful in differentiating the relative goodness of fit between different functional forms. However, comparisons of such goodness of fit, especially between models with different numbers of parameters, are not necessarily straightforward.

NOTE !!!

CHAPTER 3. FRACTAL PATTERNS EMBEDDED IN TWO DIMENSIONS

In this chapter, we will look at fractal patterns that are embedded in the plane.[1] As with other kinds of fractals, these can be strictly self-similar or statistically so. In social science, the latter are more common, but both kinds are known. In human affairs, these kinds of fractals are usually studied by geographers, who, by virtue of their specialty, focus on the analysis of spatial patterns. In consequence, they enjoy the happy distinction of having been early and enthusiastic students of fractal geometry.

> First, the concepts of a fractional dimension and dependence of measure on scale are quite foreign to much of mathematics and may therefore offer the first effective tools for understanding the irregularity widely observed in the geometry of real phenomena. Second, the numerical value of D may be the most important single parameter of an irregular cartographic feature, just as the arithmetic mean and other measures of central tendency are often used as the most characteristic parameters of a sample. (Goodchild & Mark, 1987, p. 267)

Not surprisingly, the literature on fractal analysis in geography is large. In human geography, fractals are mostly found in aspects of settlement patterns. Some geographers have even proposed a fractal principle or law in geography: "Geographic phenomena reveal more detail the more closely one looks and…this process reveals additional detail at an orderly and predictable rate" (Goodchild, 2004, p. 302). In view of the wide applicability of fractal geometry to human geography, we have chosen some of the examples in this chapter from that field.

Let's begin by thinking about what a fractal curve might look like. The classic example of a fractal curve is the Koch curve, originally conceived by the Swedish mathematician Helge von Koch in 1904 (Peitgen et al., 1992, pp. 89–93). The construction of the Koch curve is easily described: Start with a line segment. Divide it into three pieces of equal length. Take away the middle segment and replace it with an equilateral triangle. Erase the base of the triangle. This gives you a "peak" with two flanking "wings" composed of four line segments of equal length. Now repeat this process on each of the four line segments composing the new figure. Figure 3.1 illustrates the first four iterations of this process, although of course the iteration can continue indefinitely. This procedure can be described more formally as a similarity transformation involving a contraction, translation, and rotation (Peitgen

---

[1]Although fractals are often embedded in spaces of integer dimension, they can nevertheless be embedded in a space of fractional dimension, that is, within another fractal.

et al., 1992, pp. 168–169). At each iteration i, the number of line segments increases by a power of 4. We begin with $4^0 = 1$ line segments. After the first iteration ($i = 1$), we have $4^1 = 4$ segments, then $4^2 = 16$ segments, then $4^3 = 64$ segments, and then $4^4 = 256$ segments. So, the number of iterations i relates to the number of line segments $n(i)$ as 4^i. The size of the line segments decreases progressively by a factor of ⅓ at each iteration. So, the size of each of the four line segments in Step 1 is ⅓ the size of the initial segment in Step 0. Similarly, the size of each segment in Step 2 is ⅓ the size of those in Step 1, or ⅓ × ⅓ = (⅓)² = 1/9 of the original segment. In the next iteration, Step 3, the pattern repeats itself; we have segments ⅓ × ⅓ × ⅓ = (⅓)³ = 1/27th the size of the original. So, the size of the segments is related to the number of iterations by (⅓)^i. Of course, in theory, the scaling process continues ad infinitum. The mathematical object regarded as the Koch curve exists in the limit as the number of iterations goes to infinity.

Looking back at the self-similarity relation in Equation (1.2) from Chapter 1, we see that the dimension of the Koch curve can be calculated analytically as follows:

$$D = -\left(\frac{\log 4}{\log \frac{1}{3}}\right) \approx 1.26186.$$

The Koch curve can be randomized in various ways, some of which produce a curve with the same fractal dimension, whereas others generate curves with different dimensions (Batty & Longley, 1994, pp. 100–105). For example, merely by randomizing whether the added triangle points outward or inward at each iteration, one can create surprisingly irregular shapes. By varying other details of construction, many different fractal curves can also be defined along the general lines of the Koch curve. Peitgen and his colleagues (1992, p. 93) even show how a similar recursive procedure can create a honeycomb of hexagons.

The Koch curve has a number of interesting properties. Like other ideal fractals, it is an abstract construct. It is therefore perfectly self-similar and scale invariant. By rotating three copies of the Koch curve and connecting their endpoints, you can create a closed curve that is called the Koch snowflake or island (Figure 3.2). The same snowflake can be created by beginning with an equilateral triangle and applying the Koch transformation to each of its sides. What is the length of the Koch curve? The total length is the number of segments multiplied by their length. Well, it is easy to see that at each iteration, the total length increases by a factor of 4/3 because each segment is replaced by four new ones of ⅓ the size. As the number of iterations goes to infinity, so too must the length of the whole curve. In contrast, the area of the snowflake is not infinite. It can be shown that its

42

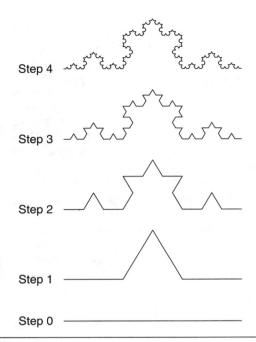

Figure 3.1 First four iterations of the Koch curve.

AMAZING!!!

area equals $\frac{2}{5}\sqrt{3}$ (Peitgen et al., 1992, pp. 147–150). Thus we have a finite area with an infinite perimeter.

Like other early fractals, the Koch curve was devised for mathematical purposes unrelated to fractal analysis, which of course did not exist at the time. Koch proposed this construct as an example of a curve that was continuous but not differentiable. It can be shown to be continuous, but in the limit, as the number of iterations goes to infinity, the curve is all corners. Because a function cannot be differentiated at a value where it forms a corner, the Koch curve is said to be "everywhere continuous, but nowhere differentiable." It is now known to be one of a large class of such functions. Nowhere differentiable functions are not mere spoilers designed to create paradoxes in calculus. True fractals are characteristically nondifferentiable because of their irregularity and discontinuity.

The Koch curve is very pretty and certainly curious, but does it have any relevance to human affairs? In fact, it does, in several ways. For example, Batty and Longley's (1994, pp. 165–195) measurement of the urban boundary

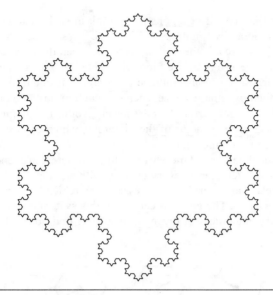

Figure 3.2   The Koch snowflake.

of Cardiff finds a fractal dimension similar to that of a Koch curve. Does the dynamic recursive construction of the Koch curve provide a model for a city boundary? The connection is not obvious, but it might provide a useful heuristic. In fact, though, other models have been proposed to explain the fractal outlines of cities.

Central place theory provides us with a theoretical example of a strictly self-similar fractal from economic geography that relates to the Koch curve. Christaller (1933/1966) developed the theory of central places to explain the irregular distribution of settlement types and sizes. Lösch (1954) significantly elaborated the theory and mathematics, providing a more general theory with a clearer derivation. The classic formulation of the central place model predicts that on an unbounded, featureless plane with homogeneously distributed resources, central places (centers of population, manufacturing, administration) will dominate hexagonal regions tiling the plane. Major centers are surrounded by six second-order centers arranged hexagonally, and each of those centers is in turn surrounded by six smaller centers, and so forth, down to the level of the agricultural hamlet. The model predicts the locations of different types and sizes of settlements that maximize the efficiency of production, marketing, redistribution, and administration under different conditions or assumptions. The geometric result is a series

44

of overlapping lattices of hexagons that shrink in scale with each level of the settlement hierarchy (Figure 3.3). The three classic cases studied are for $k = 3, 4,$ and $7,$ where k is the scaling factor between the areas of the hexagons in successive levels of the hierarchy. Thus, in a $k = 3$ central place system, the largest hexagons embrace an area that is equal to three of the hexagons of the next smallest size, each of which in turn contains the area of three of the hexagons from the next lower level of the hierarchy, and so forth. The linear scaling factor of the hierarchy, which corresponds to the distance between settlements, is $\sqrt{k}$.

Central place theory is not merely of historical interest in economic geography. "Studies undertaken in every country with a complex modern economy show retail and service businesses hierarchically organized in a central-place system. The general features of this system repeat themselves in each country, regardless of divergent historical traditions" (Berry, 1967,

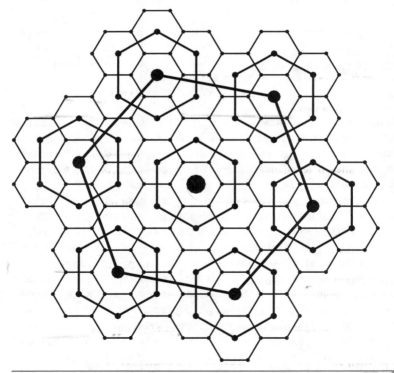

Figure 3.3 Diagram of a 4-tiered hexagonal central place lattice for $k = 3$. The sizes of the dots and the thickness of the lines reflect the levels of the hierarchy; that is, the largest central place is shown by the largest dot.

p. 89). A substantial literature on it has grown up (see Mulligan, 1984, for a review). Many extensions of it have been developed that study its dynamics under varying assumptions. It continues to inspire further research (e.g., Krugman, 1996, pp. 88–92; Peng, 2004; South & Boots, 1999). A review of the many variations of the theory is beyond the scope of this monograph; here we wish to briefly discuss its geometry.

Sandra Arlinghaus (1985, 1993) has shown that the hexagonal central place lattices are ideal fractals that can be constructed through a Koch-type similarity transformation. Let's begin with the simplest case, the $k = 3$ hierarchy. Start with a hexagon. Replace the first edge with a wedge composed of two line segments of equal length having an included angle of $120°$. As with the Koch curve, the wedge should point outward. Replace the next edge with the same object, but point the wedge inward. Continue around the hexagon, replacing the straight edges with the wedges, but alternating their direction. The alternation conserves the area of the original hexagon. Iterate the similarity transformation, and it reproduces the nested hexagonal lattices of a $k = 3$ central place hierarchy. The dimension of the fractal can be solved analytically as the number of pieces N increases by 2 and the scaling factor is $\sqrt{3}$; thus, $D = 1.26186\ldots$, the same as that of the classic Koch curve. For a $k = 4$ lattice, $D = 1.585\ldots$, the value of the fractal dimension of the Sierpinski triangle, and for a $k = 7$ hierarchy, $D = 1.129\ldots$, which is also the value of a variant of the Koch snowflake called the "Gosper Island" (Mandelbrot, 1983, pp. 46–47).

Thus, the Koch snowflake is not just a pretty picture or a mathematical curiosity. It actually has practical implications for interpretation in social science because it is related to patterns from both theoretical economic geography and empirical observation. For example, the new economic geographers, who attempt to model the self-organizing dynamics underlying the evolution of spatial economic behavior (e.g., Fujita, Krugman, & Venables, 1999; Krugman, 1996), need to ensure that their models reproduce such fractal patterns. This may be challenging, just as it has been difficult to create spatial economic models that reproduce the observed power law distribution of cities (Fujita et al., 1999, pp. 215–225). In the next section, we will study how to measure empirical fractals embedded in two dimensions. To build on what we have already learned, we will first consider a simple example that is analogous to the Koch curve, a coastline.

Estimating the Fractal Dimension of Empirical Data

Several methods of estimating the fractal dimension of empirical data sets have been developed. Here we will discuss two, the divider method and the box-counting method, although others exist as well. The different methods are typically used under different circumstances. Each has its unique advantages and disadvantages.

The Divider Method: How Long Is the Coast of Britain?

Most coastlines, contour lines, and rivers are fractal curves. Mandelbrot's famous article "How Long Is the Coast of Britain? Statistical Self-Similarity and Fractional Dimension" (1967) offers a classic illustration of this kind of curve and describes a method of measuring it. This is the so-called divider method. Dividers are instruments that are used for measuring in drafting and navigation. A pair of dividers looks much like a compass, but instead of having one leg with a point and the other with a pencil, dividers have two legs with pointy ends (Figure 3.4).

Imagine a large and detailed map of a coastline. We can use dividers to measure the length of the coast by "walking" the dividers along the coastline. Start with the dividers set to a large setting, say, 100 km, and then walk the dividers down the irregular line formed by the coast. The total length of the coastline measured in this way will be equal to the number of "steps" taken by the dividers, multiplied by their setting (i.e., 100 km). Much of the detail of the coastline, small bays and inlets and narrow promontories, will, however, be skipped over by the path of the dividers. Then reset the dividers to a smaller setting, say, 10 km, and repeat the measuring process. You may expect the coastline to have the same length as before, but it won't. In fact, the result will be very different: the total length of the coastline will be longer—in fact, much longer—because much more detail will have been measured with the smaller divider setting. Many of the inlets and peninsulas previously skipped over will now be measured with the dividers partly closed. Try this yourself. Repeat the measuring procedure with the dividers set to 1 km, 100 m, 10 m, and so forth. The measured length of the coast will grow rapidly and ultimately approach infinity as the unit of measurement (i.e., the divider setting) approaches zero. What is the real length of the coast? This is indefinable because it depends on the scale of measurement. In an important sense, one cannot speak of the true length of a river or a coastline; one can specify only how irregular it is. The fractal dimension quantifies how irregular and complex the curve is, and it is easily calculated from the data collected with the dividers.

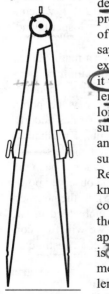

Figure 3.4
Drawing of a
pair of dividers.

The measurements of the coastline taken with the dividers can be described with a simple power law relation that reflects the dramatic increase in the measured length of coast as the size of the measuring unit decreases. The total length of the coast's curve L is related to the length of the unit of measurement r according to the following function:

$$L(r) = Kr^b, \qquad (3.1)$$

where K is a positive constant of proportionality. We really wish to calculate the exponent b because it is related to the fractal dimension. To calculate b, we take the logarithms of the step length (divider setting) and plot them against the logarithms of the total length of the coastline as measured at each setting. As usual, the base of the logarithms is unimportant, provided they are the same, because the result will be the ratio of two logarithms, and their ratio will be the same regardless of the base. The function represented in the scatterplot will be linear if the coastline is a fractal curve. If the relation in the scatterplot is not linear, then the curve that was measured is not fractal (or at least not a simple fractal). If it is linear, draw a best-fit line through the data points. The slope of the line is an empirical estimate of b. In Equation 3.1, the exponent $b = 1 - D$, where D is the fractal dimension. Here's why: For any particular measuring unit, the length of the curve equals the size of the "meter stick" (divider setting) r multiplied by the number of steps N walked down the curve ($L = N \times r$). The number of steps in turn depends on the size of the measuring unit according to the self-similar relation $N = Kr^{-D}$. Substituting Kr^{-D} for N in the previous equation, we obtain $L = Kr^{-D}r^1 = Kr^{1-D}$. AHA!!!

The dimension D tells us how irregular and complex the curve of the coastline is. For a straight coastline, or any straight line, it will equal 1 and it will increase with the complexity of the coastline until, at $D = 2$, the line actually fills the plane. Mandelbrot (1967) suggested that most coastlines could be modeled by the Koch curve because the dimensions of some coastlines are close to that of the Koch curve, $D \approx 1.26$. Geomorphologists have studied this question and marshaled data to support this argument (e.g., Lam & De Cola, 1993, pp. 24–29; Tanner, Perfect, & Kelley, 2006; Turcotte & Huang, 1995, pp. 14–17). The same concepts also apply to many fractal curves, including most coastlines, lakeshores, topographic contour lines, river plan views, and outlines of vegetation patches. These are, of course, natural phenomena, but we can also include in the list social phenomena such as the outlines of cities (Batty & Longley, 1994, pp. 165–195), which can possess boundaries with similar dimensions; political boundaries that follow natural features; and the highly irregular boundaries of some political units, such as gerrymandered voting districts.

As one would hope, this method will not yield fractal results for Euclidean figures; they will yield integer, not fractional, dimensions. If one attempts to use the divider method to measure the fractal dimension of a circle, the total perimeter measured quickly converges to a limit representing the real length and the log-log plot approaches a horizontal line. This means that $b = 0$ and so $D = 1$—an integer, not a fraction. So, this method of analysis also gives us the correct dimension for Euclidean figures.

Although the walking divider method has been used extensively, it is not appropriate for all kinds of fractals. The method cannot be used with overlapping or discontinuous lines, or with most phenomena of more than two Euclidean dimensions. In addition, questions have been raised about its accuracy and precision under certain common circumstances, such as finite or truncated data sets or data with a periodic trend. For example, there has been some discussion of how to handle the remainder when the curve being measured is not an integral multiple of the length of the measuring unit. Various modifications of the method have been suggested to ameliorate the problems (e.g., Batty & Longley, 1994, pp. 172–195; Nams, 2006; Rice-Snow & Emert, 2002), but other methods of estimating fractal dimension are often preferred.

The Box-Counting Method

Box counting is probably the method best known and most commonly used to analyze sets embedded in two dimensions. It is a versatile method because, unlike the divider method, it can be used with any kind of set embedded in two dimensions. It is also easily generalized to other embedding dimensions, which enhances its popularity. The basic concept is closely related to the capacity and Hausdorff dimensions, but instead of creating a covering of the set using disks, balls, or spheres, you create coverings using squares.

As in the other methods we've reviewed, box counting is designed to simultaneously allow you to determine whether a pattern is fractal and, if it is, to estimate its fractal dimension. Normally, you would start with some complicated-looking set or object that you think might possess self-similar or scale-invariant qualities. It could be a curve, such as the outline of a city or a voting district, or a collection of such curves. It might be a pattern composed of a map of streets or roads or railroads. It could even be a point pattern, such as the locations of houses or settlements or stores. We've used this technique to study the pattern of buildings with in ancient Mayan cities, the patterning of archaeological sites found during archaeological surveys, and the spatial distribution of artifacts on ancient living floors uncovered at archaeological sites. Whatever its substantive content, we would expect the pattern to look fairly complex. We would then use the box-counting method to address the questions "Is this pattern best described as a fractal?" and "If so, what is its dimension?"

This is the procedure: You overlay a grid of squares on the object to be measured, and then you count the number of boxes that contain part of the design, that is, intercept at least one point of the set. The number of squares N required to cover the set will depend on the linear size of the squares, r, so N is a function of r, or we can write $N(r)$. Now you reduce the mesh size of the grid and again count the number of boxes occupied by at least some part of the object. The number of occupied boxes will, of course, increase because the boxes are smaller. If, however, the pattern is a fractal, the number of occupied boxes will rise dramatically, more than you would expect from the linear change in the size of the boxes alone. As with the divider method, this occurs because as you zoom in on or magnify a fractal pattern, more detail is revealed. As you shrink the size of the boxes, you are zooming in on the figure and capturing more detail. So, you repeatedly reduce the size of the boxes in the grid overlaid on the pattern, each time recording the two variables, N and r. Next, you plot the log of $N(r)$ against the log of r. If the relation between the two is linear (it is a power law) and therefore the pattern is fractal because, as we know, the self-similarity of a fractal produces more much detail as you measure it at ever finer scales. If we call the slope of the best-fit line on our plot b, then the fractal dimension is $D = -b$. Let's look first at a simple nonfractal, Euclidean example to illustrate how box counting can measure dimension.

The area A of a circle is related to its radius r by a simple power law known to all high school students; $A = \pi r^2$. Thus, if we increase the radius by a certain amount, then the area increases by a power of two times the constant, π. Now, the area of a circle can be approximated by a covering of squares. So, if we cover a circle with 4 squares, for instance, as we show in Figure 3.5, and then reduce the linear size of the squares by ½ while holding the radius of the circle constant, the number of squares required for the covering increases to 16 (Hastings & Sugihara, 1993, pp. 37–38). Analytically, if we scale the boxes down linearly by ½, we get 4 times as many boxes. $\mathrm{Log}(4)/\mathrm{Log}(\tfrac{1}{2}) = -2$ and therefore D equals 2, the true dimension of the circle, which is, of course, not fractal.

Let's look now at a random fractal example. The fractal character of transport systems has been studied by many geographers. Not surprisingly, most studies are of road systems, which are the salient patterns that emerge from our settlements, above all in auto-intensive societies like the United States (e.g. Burnett & Pongou, 2006; Lu & Tang, 2004; Rodin & Rodina, 2000; see also studies cited in Batty & Longley, 1994). Some rail systems

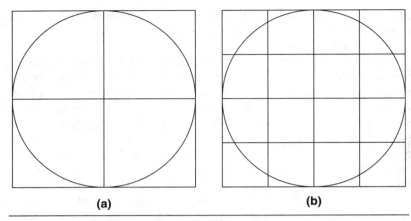

Figure 3.5 Box-counting dimension of a circle. (a) A small circle is covered by four squares. (b) We shrink the size of the squares by ½ while retaining the size of the original circle from (a), and it requires 16 squares to cover the circle (redrawn and modified from Hastings & Sugihara, 1993, Figure 3.1).

have been studied, but those were mainly urban metro systems or suburban commuter systems (e.g., Benguigui, 1995; Benguigui & Daoud, 1991). In contrast, few, if any, studies have looked at interurban or long-distance passenger and cargo rail networks. In Figure 3.6, we show a map of the rail system in the region surrounding Albany and Schenectady, in New York State, drawn from the U.S. Census Bureau's TIGER (Topographically Integrated Geographic Encoding and Referencing) line files. It looks irregular, but is it fractal? We performed box counting by laying a series of four grids over the image, as shown in the figure. At each iteration of the procedure, we shrank the linear sizes of the boxes in the grid by half and counted the number of boxes that included any part of the image. The sizes of the grid squares (the lengths of the box sides) and the number of boxes occupied by the data set are presented in Table 3.1.

We then plot the logarithms of the box sizes against the logarithm of the number of occupied boxes (Figure 3.7). Inspection of the graph reveals that the relation is strongly linear. This implies that the relation between the scale of measurement and the amount of detail revealed is a power law and thus scale invariant—in short, a fractal pattern. The slope of the best-fit line is about −1.42, and so the box-counting estimate of the fractal dimension is $D = 1.42$. Our result is close to that of Benguigui and Daoud (1991) for the Paris rail system ($D = 1.466$), but please note that we do not consider ours to be a serious research finding, which would obviously require much more extensive and detailed analysis. We are merely illustrating the technique with a convenient data set.

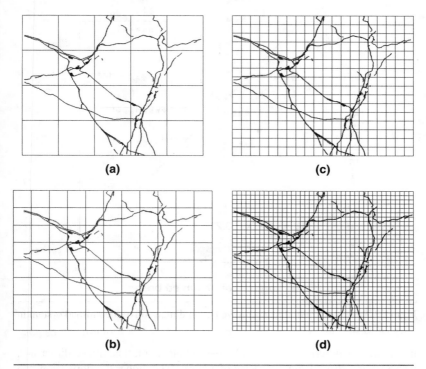

Figure 3.6 Box-counting procedure illustrated for a map of the Albany, New York, regional rail system.

Box size (r)	Number of occupied boxes N(r)	Log (r)	Log (N(r))
25	16	1.397940009	1.204119983
12.5	48	1.096910013	1.681241237
6.25	125	0.795880017	2.096910013
3.125	308	0.494850022	2.488550717

Table 3.1 Box-counting statistics for the Albany Area Rail Network.

What might our result mean? First and foremost, it is an accurate and meaningful measurement of the phenomenon. This is not trivial. Proper description precedes explanation, as the history of science amply demonstrates. More

52

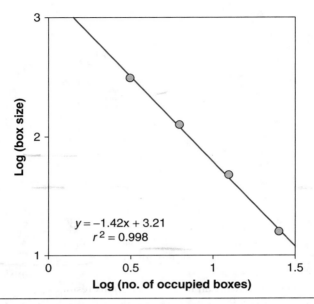

Figure 3.7 Log-log plot illustrating the fractal dimension of the railroad system in the vicinity of Albany, New York.

important, the description can lead us to an explanation. The fractal dimension measures the space-filling property of a curve. The higher the dimension, the more the curve fills the plane. So, a rail network with a higher dimension comes closer to more points in the plane than one with a lower dimension. Therefore, a high dimensional network can probably get you, or your cargo, closer to its destination than a low dimensional one. Not surprisingly, road networks tend to have higher dimensions than rail networks. And, older and larger cities tend to have higher fractal dimensions than younger or smaller ones, facts that clearly tell us something important about urban growth processes (e.g., Batty, 2008; Batty & Longley, 1994; Benguigui, Czamanski, Marinov, & Portugali, 2000; Lu & Tang, 2004). The deeper lesson is that to understand this type of phenomenon, you must first be able to measure it, and the measurement can lead you to theories about its dynamic properties, as it has with cities and transportation networks.

 The parallels between the box-counting procedure and the capacity dimension are extremely close. The box counting is a type of covering, although instead of employing disks or balls, it uses squares, which considerably facilitates its implementation. Obviously, the relationship between a covering of disks and a tiling of boxes can be studied both analytically and empirically, and such analyses have shown that box counting normally produces the same result as the capacity dimension (Hastings & Sugihara, 1993, pp. 43–44). Numerous studies support its accuracy and precision under common conditions and reasonable circumstances. Unfortunately, the box-counting dimension is not always exactly the same as the Hausdorff dimension (Falconer, 2003; Peitgen et al., 1992, p. 219), but this is rarely an issue with typical empirical applications.

Applying the Method in Practice

To avoid naive mistakes, one should consider several practical issues when implementing the box-counting method. To begin with, to estimate the dimension accurately, one should minimize the number of occupied boxes, $N(r)$, at each stage of the calculation. Deviations from this rule contribute small errors to the estimate of dimension. One solution with random fractals is to offset the grid repeatedly, perhaps randomly, to minimize $N(r)$ at each stage of the analysis. An alternative is to rotate the grid. It is also important to fashion a grid that tightly matches the boundaries of the object that you wish to measure. If you use a mesh or grid that is arbitrarily larger than the fractal itself, you can introduce spurious results.

You should also seek to use as large a range of box sizes as possible, but within the finite limits of the data set. Obviously, box sizes that are larger than the entire data set or smaller than a single data point do not contribute additional information to the analysis and should be avoided. Empirical random fractals typically exhibit fractal behavior within some range of scales, and it is desirable to identify that range. One normally expects the range of fractal behavior to extend over at least a couple of orders of magnitude. If the fractal property extends over only a very short range, its fractal dimension may not be a meaningful description of the data. (Note, however, that a single phenomenon can exhibit multiple fractal dimensions, each extending over its own range of scales.)

The increment by which you reduce the box sizes will have an effect on the analysis (Foroutan-pour, Dutilleul, & Smith, 1999) because each box size becomes a case in the subsequent analysis of the power law relation. For ease of analysis, it may be desirable to ensure that the logarithms of the box sizes are evenly spaced on your log-log plot. It is also preferable to perform the analysis at many different box sizes because each iteration of the process contributes an additional data point to the analysis on your log-log plot. If the increment of reduction is large, then you cannot use many box sizes because you quickly reach the finite limits of the fractal. On the other hand, if you choose too small an increment, then the number of occupied boxes may not change between iterations, creating a series of undesirable plateaus in the graph. Bear in mind as well that the choice of

54

box sizes is analogous to the issue of histogram binning discussed in the previous chapter.

As we have seen, fractal patterns are ubiquitous in geographic data. If analyzing geographic data, you should consider using a geographic information system (GIS). GIS applications, such as ArcGIS (published by ESRI), are designed to manipulate geographic data correctly and easily, which obviously facilitates analysis. For example, many geographic data sets are available in geographic coordinates (i.e., latitude and longitude) in decimal degrees, whereas box counting probably ought to be performed in the plane. Therefore, the original data need to be projected, which a GIS will do with a few simple commands. Of course, one should select a projection with care because different projections conserve different properties of the original data. Therefore, the choice should take into account the fractal aspect of the phenomenon that you are studying. In principle, it is not difficult to implement box counting in a GIS computing environment by writing an appropriate script, and many investigators have done so (e.g., Bren, 1995; Crawford, Commito, & Borowik, 2006; Shen, 2002).

When fractal analysis began in earnest in the 1980s, box counting (along with other fractal analysis methods) was found to be so computationally intensive that it presented problems to the mini- and microcomputers that were then becoming common. Motivated by this problem, Liebovitch and Tóth (1989) developed a computationally fast algorithm for implementing box counting. The approach they proposed has been widely used and implemented (e.g., Block, von Bloh, & Schellnhuber, 1990; Buczkowski, Kyriacos, Nekka, & Cartilier, 1998; Kruger, 1996; Molteno, 1993). The power, speed, and memory of current computers have reduced, but not eliminated, the computational barriers to fractal analysis. If you are analyzing large data sets and the processing time seems prohibitive, you may wish to consult the references just cited.

Finally, box counting is easy to generalize to data embedded in other dimensions. It should be easy to visualize how box counting can become "cube counting" for sets embedded in three dimensions. Although it is difficult to visualize, the same procedure works for sets embedded in even higher dimensions. It can also be used for sets embedded in one dimension, such as the Cantor set.

Summary and Discussion

In this chapter, we have discussed fractal patterns embedded in two dimensions. We looked at several examples of such fractals, including both exactly self-similar ones and stochastic ones. We explained two methods of estimating the fractal dimension for such data sets, the walking divider method and the box-counting method. Other methods also exist, but space constrains our ability to review them here.

Human geography seems to offer the largest number of examples to social scientists of fractal patterns embedded in two dimensions. Geographers have developed a substantial and coherent literature that employs fractal analysis to describe and measure spatial phenomena as well as develop dynamical theories and models of economic and social behavior.

We do not wish to leave the impression that the only fractal patterns in this category are geographic. Works of art have also been analyzed using fractal techniques. The best known examples are probably Richard Taylor's fractal analyses of Jackson Pollock's drip paintings (Taylor, Micolich, & Jonas, 1999, 2002), but Voss (1998) has analyzed Chinese landscape paintings and Brown et al. (2005, p. 57) have examined the figurative painting on an ancient Mayan funerary vase. We can also point to the analysis of Bender Gestalt drawings by Sarraillé and Myers (1994). In sum, fractal analysis has been applied to a variety of different patterns related to human behavior.

CHAPTER 4. SOCIAL PROCESSES
THAT GENERATE FRACTALS

In the preceding chapters, we have emphasized that the data from many different examples in the social sciences are not fit well by the classical "normal" distribution in which the values cluster around a single typical value that is characterized nicely by the mean. We have described how these data are better fit by fractal descriptors that span many different scales of measurement and are described well by the fractal dimension, which is a quantitative measure of how the data depend on the scale at which they are measured.

It is natural to ask what processes lead to such fractal distributions in data from the social sciences. Or, stated in another way, when we observe a fractal distribution, how does that inform us about the relevant social mechanisms that generated the data? In the physical sciences, many different mechanisms can produce similar fractal patterns. It is straightforward to derive the fractal data that would be produced by a unique physical mechanism. However, the "inverse problem"—figuring out the physical mechanism from the fractal data—can be frustratingly difficult because different mechanisms can produce the same fractal data set, a phenomenon sometimes called *equifinality*. In other words, the answer to the question "What process produced this fractal data set?" is not necessarily unique; therefore, inferring the dynamical process from the topology of the fractal data set is problematic. The situation is similar in the social sciences as well. The fractal data alone cannot tell us the unique social mechanism that produced them. However, they can give us quite interesting clues. In a negative sense, they let us rule out social mechanisms that do not produce fractal data. In a positive sense, our previous experience with the known mechanisms that produce fractal data help us frame hypotheses to search for the causes of the observed data.

In this chapter, we provide some examples of mechanisms that have been proposed to explain fractal data observed in the social sciences. Rather than a comprehensive compendium, this brief bestiary of mechanisms will provide a few relevant examples that serve as specific mechanisms, or analogies of mechanisms, or the starting point of creative new thoughts about mechanisms, to help the reader form hypotheses about the processes responsible for other fractal patterns that may be observed in the social sciences.

How We Do Our "To Do" Lists: Preferential Priorities

Apparently, the times between answering our e-mails, making telephone calls, playing games, or trading currency futures are not distributed in a

56

uniform way. There are bursts of rapidly occurring events separated by ever longer periods of inactivity. The distribution of times between these events is not a normal distribution, but rather a fractal power law distribution, where the probability of the time t between events has the form t^{-a}, where a is typically about 1. Why is this?

We keep adding items to our "to do" list. Let's say that we give each item equal priority and then choose an item, at random, to do. This leads to an exponential distribution, e^{-bt}, which is very far from the observed power law t^{-a}.

Barabási (2005) proposed that the power law distribution of the times between events that we do "can be explained by a simple hypothesis: humans execute their tasks based on some perceived priority, setting up queues that generate very uneven waiting times distributions for different tasks" (p. 211). The mathematical form of his model is that we have a "to do" list of L tasks, each of which has a priority chosen from a distribution of probabilities $r(x)$. We do the task with the highest priority on the list and then replace it with another task with a priority chosen from $r(x)$. If that is the case, then the probability that any task spends time t on the list has the fractal power law form t^{-1}. Interestingly, he also showed that this waiting time distribution does not depend much on the probability distribution $r(x)$ of how a person sets his or her priorities.

The predictions of this quite simple and elegant mechanism match the observed fractal data, but others have suggested other possible mechanisms such as circadian rhythms and other periodic phenomena (Malmgren, Stouffer, Motter, & Amaral, 2008). What experiments could you devise that would test the validity of this explanation? How can you use this understanding to better allocate resources such as pricing plans for telephone companies, corporate inventory management, or human foraging patterns? Can you generalize this mechanism to explain other observed fractal data, such as the bursts of creative ideas that emerge together in social systems?

How We Kill: Attendant Causes, Self-Organized Criticality, and Agent-Based Models

Murderers, gangs, rioters, and armies kill people. We're quite a species. Richardson, a British Quaker who served as an ambulance driver in World War I, wanted to know how often we kill how many others, as a route to understanding why we do so. He is best known in the field of fractals for his analysis, popularized by Mandelbrot (1967), showing that the length of the west coast of Britain depends on the scale at which we measure it, because finer scales include ever more, ever smaller bays and peninsulas. Richardson (1941, 1948, 1960) found that the number of persons who died because of a deadly quarrel that "includes murders, banditries, mutanies, insurrections, and wars" (Richardson, 1960, p. 6) is a fractal power law. Ever more deadly killing events are ever more rare. How often N people died is proportional to N^{-b}, where b is between 1 and 2. A similar fractal power law scaling has been found by others as well, both for the number of battle deaths and the fraction of battle deaths compared to the whole population (Roberts & Turcotte, 1998). What do these notable relationships tell us about the causes of wars or how people behave in them?

Richardson (1948) himself noted that the distribution of the size of cities, the number of people in gangs in Chicago, and the number of raids by bandits in Manchuria also had power law fractal distributions, with the same scaling exponent b as the one he found in the deadly quarrels. Are the statistics of these deadly quarrels therefore simply a reflection of the statistics present in these other social phenomena? In what other social systems do you think the existence of a fractal power law distribution in one variable produces a similar distribution in another dependent variable?

Roberts and Turcotte (1998) suggested a different mechanism. A model called self-organized criticality (SOC), originally developed in the physical sciences, has been useful in understanding many different phenomena, including earthquakes, epidemics, forest fires, traffic jams, city growth, market fluctuations, and firm sizes (Bak, 1997; Cederman, 2003; Turcotte & Rundle, 2002). In all of these systems, stress that exceeds a local threshold is then spread to neighboring points. Thus, everywhere, the local stress adjusts to being just under this threshold. The system is poised at its most unstable state—just a tiny bit more and the stress is shared—but each local point lives just shy of that threshold of instability. This is a very different picture from our usual thinking, where we assume that a system is gently resting at its most stable equilibrium. The eponymous example is that of a sand pile, where grains of sand are slowly added to the pile. The local slope of the sand at any one point increases as new grains of sand reach it until it exceeds the stable threshold, and then it sheds its grains downhill, where the local slope may then exceed its threshold of stability, pushing the avalanche further downhill. At every point, the slope of the sand settles into an angle just below that of the threshold needed to cascade downhill. If the slope were steeper, the sand would flow. It is called a "critical" model because the slope of the sand at each point is poised between remaining stable and starting an avalanche, just as water at the temperature of freezing is poised between the "critical" transition between a liquid and a solid. It is called "self-organized" because the action of the sand itself brings the system to this critical point; no magic controller from the outside is needed. The avalanches of sand are fractal in mass, space, and time. There are power law distributions of the amount of sand in the avalanches, in the area over which the avalanches travel, and in the timing between the avalanches. Roberts and Turcotte (1998) proposed that countries are poised for war just like the sand in the sand pile. When an event triggers the war, it spreads like the avalanche of sand, most often engaging only a few countries, but sometimes many more countries, reproducing the power law fractal distribution of deaths. How can these mechanisms be generalized to other social phenomena? What stresses are shed from one person, social group, company, or nation-state to another that could organize the distribution of a measured variable into a fractal power law distribution?

Cederman (2003) writes that "the mere presence of power laws does not guarantee that the underlying process is characterized by SOC....Such accounts ultimately hinge on the theoretical and empirical plausibility of the relevant causal mechanisms" (p. 137). He proposes a model in which each nation acts as its own autonomous agent, following its own few simple rules. This is a class of models called agent-based models. It has been quite surprising how unanticipated collective behavior emerges from a collection of such agents following simple similar (or dissimilar) rules. His model reproduces the fractal power law distribution of deaths and its scaling exponent b through the following rules: (a) Each nation's resources depend on its area; (b) each nation allocates its resources to face the threats at its fronts from its neighbors; (c) each nation decides, with a probabilistic criterion, to attack each of its neighbors; (d) the success or failure of these attacks alters the areas of each nation; and (e) these processes repeat at the next time step. Is such a model too incomplete in describing international relationships or too complex using more rules than are necessary to reproduce the fractal power law of deaths? How do we decide how many rules are needed in an agent-based model? How do we generalize such models to other social situations? Do we already need to know the answer from previous experimental studies or our own intuition before we can develop such models?

How We Network: Preferential Attachment

John Donne said that "no man is an island"; we influence and are influenced by the people who surround us. The first step in understanding the role of these social networks in shaping us is to understand the structure of these networks. Quantitative analysis of many social networks, such as those of articles that are cited in other publications, pages connected by links (URLs) on the Internet, actors together in the same movies, and scientists who publish together as co-authors, all have a similar structure (Jeong, Neda, & Barabási, 2003). In these networks, the degree distribution—the number of nodes (such as articles, Web pages, actors, or authors) that have links to k other nodes—has a fractal power law distribution proportional to k^{-c}, where c is a number between 1 and 4.

60

These social networks typically grow larger in time as new nodes and links are added. Linking the new nodes at random to the existing nodes does not produce a network with a fractal power law degree distribution. Barabási and Albert (1999) and Pastor-Satorras, Vazquez, and Vespignani (2001) proposed that these networks grow by linking new nodes preferentially to the nodes that already have the largest number of links. As Barabási (2002) explains,

> When choosing between two [Web] pages, one with twice as many links [into it] as the other, about twice as many people link to the more connected page.... Preferential attachment rules in Hollywood as well... the more movies an actor has made, the more likely it is that he or she will appear again on the casting director's radar screen. (p. 85)

When the preferential attachment to a node occurs with probability proportional to the number of existing links at that node, then the network develops a fractal power law degree distribution. Does this type of preferential attachment explain the structure of compadrazgo networks or marriage systems? What other ways are there to define the preferences for adding a new node and its links, and what patterns in the structure of networks would they produce?

How We Decide Where to Live: Diffusion Limited Aggregation

The buildings and people in a city and its surrounding countryside sprawl over the landscape. But the patterns they form are not strictly regular or uniformly random. There are ever larger centers of concentration in local neighborhoods, districts, cities, counties, states, and nations forming hierarchies of clusters of different sizes. The pattern is a self-similar fractal; the smaller clusters are reproduced at ever larger scales (Batty & Longley, 1994). The mass, M, of people within any given radius, r, has the fractal power law scaling that M is proportional to r^d, where the fractal dimension d is in between 1.4 and 1.9 (Sambrook & Voss, 2001). Why do people form such patterns?

Geographers have asked whether the models that produce similar patterns in physical systems can lead us to the analogous social mechanisms that produce those patterns among people. For example, these physical patterns are produced when a light slippery fluid is pushed into a thick viscous fluid; when the spark of an electric current tears its way through an insulating medium; or when ions in an electrochemical reaction build their branched, feathered filigree on an electrode. Although the flow of fluids, electricity, and ions are all governed by different physical laws, all of these laws have the common feature that the growth is fastest at the tips of the pattern because new material is added proportional to the local gradient of the existing structure (Liebovitch & Shehadeh, 2003). This mechanism of growth, and the patterns it produces, can be modeled by a process called diffusion limited aggregation, or DLA (Meakin, 1986). Batty and Longley (1994) proposed that cities grow in an analogous way as people settle next to already established areas. If people are only partially influenced by the established areas, then their probability to settle, p, determines the urban pattern and its fractal dimension (Batty & Longley, 1994; Meakin, 1986). When p approaches 1, people form a solid dense cluster with no open spaces that has fractal dimension 2. When p approaches zero, people form a delicate tree pattern with many open spaces of all different sizes that has fractal dimension 1.7. What other social processes depend on new emotional, psychological, political, or economic links being more likely to be attached to the gradients of existing structures?

How We Look for Food: Lévy Flights

How does a bird efficiently search a wide ocean for scarce food? Viswanathan et al. (1996) attached electronic recording devices "to the legs of five different adult albatrosses" to measure the duration of flights between "their alighting on water to eat or rest" (p. 413). The distribution of these flights was not a uniform or a normal distribution. Instead, these flight times were a fractal power law distribution, called a Lévy flight, in which the number of intervals of flight time t was proportional to t^{-e}, where $e = 2$. Most of the time, the flights are brief, but ever more infrequently, the albatross takes an ever longer flight. They also found that the durations of separate flights were not independent of each other, as one would expect from a random walk model based on Brownian motion, but rather were correlated over short, medium, and long time scales (see Chapter 5 for a discussion of Brownian motion and fractal time series). p. 64

Why does the albatross choose this distribution of durations between its dips in the ocean? The Lévy fractal search pattern may reflect the fractal distribution of food on the ocean (Viswanathan et al., 1996), or it may be that this search pattern is the most efficient way to find food "when the target sites are sparse and can be visited any number of times" (Viswanathan et al., 1999, p. 911). But science, like life, is often not so simple. Edwards et al. (2007) reanalyzed the albatross data to find that the flight distributions are not a Lévy flight but a combination of two other distributions (a gamma distribution over brief flights and an exponential distribution over long flights). What about the search patterns of other animals? Edwards et al.

(2007) found that the foraging times of unfenced deer and bumblebees were not Lévy flights, as first thought. On the other hand, Sims et al. (2008) found that the foraging times of the basking shark, bigeye tuna, Atlantic cod, leatherback turtle, and Magellanic penguin are Lévy flights, and Brown, Liebovitch, and Glendon (2007) found that nomadic people also visit their camps using Lévy flights. Stay tuned, we're still arguing this one.

Would these Lévy flights be useful in finding land mines or archaeological artifacts at a site, if these targets were distributed in a regular geometric pattern, a uniformly random pattern, or a fractal pattern? What other types of random walks can you think of, and how could they be applied to finding targets in two or three dimensions, or a time series of one dimension?

How We Live Together: Balancing Cohesive and Disruptive Forces

Hamilton et al. (2007) analyzed how the number of individuals in 339 hunter-gatherer societies depends on the social hierarchy of the group. They found that a constant ratio of group sizes extends across individuals, families, extended family groups, residential groups, multiyear aggregations, and the total region population size. Each higher group level had about four times as many people as that of the groups in the next lower level. At the family level, this is understandable because it means that each family, with two adults and two children, will be at the replacement rate needed to maintain the population. But why should the same ratios hold for higher levels of organization?

They proposed that "such scaling ratios are a fundamental structural component of human social organization" (Hamilton et al., 2007, p. 2200), that the size of the groups at each level is governed by "two basic kinds of forces: (i) cohesive forces that tend to draw and hold individuals together and (ii) disruptive forces that tend to pull individuals apart and to create barriers to exchanges between them" (p. 2200). Why do the different social mechanisms between each level all act in the same way? Are there simple rules of the efficiencies to be gained (or lost) by association that are the same at each level? What other basic rules of social interactions, regardless of the specific type of social interaction, are present in our hierarchical organizations, and how would they be discoverable from experimental data?

Summary and Discussion

In this chapter, we have described several dynamic processes that produce fractal patterns of different kinds. As we mentioned at the beginning, this is by no means a complete catalog. Many other such processes and mechanisms have been proposed. Like the ones described here, they generally produce random fractals. In both nature and culture, random fractals seem more common than deterministic ones, so the processes that produce them are the ones in which we are most interested. Nevertheless, humans can create deterministic fractals, and, unlike in nature, they can be attributed to intentional, purposeful behavior. Such fractals can arise in organizational structures, art, and architecture. Of course, these deterministic fractals, like other empirical ones, have finite limits.

You may be surprised to note the absence of a discussion of chaos in this chapter. Much has been made of the relationship between fractals and chaos. Chaos is a characteristic of dissipative dynamical systems. Chaotic dynamical systems are, by definition, those that exhibit extreme sensitivity to initial conditions. These kinds of systems have a number of interesting characteristics. For example, they are unpredictable even when fully deterministic. They also produce highly complex and irregular behavior that is neither random nor periodic. This is where the relationship to fractals comes in. The solution set of a dynamical system is generically called an attractor, and the attractor of a chaotic system is specifically called a strange attractor. Strange attractors have fractal structure, and fractals are strange attractors. This has led some investigators to infer that chaotic social dynamics underlie the fractal patterns in society. Although the conclusion may be true, the inference is problematic, for two reasons. First, despite its theoretical importance, chaos has been difficult to identify in practice. Few real, empirical examples of chaotic dynamics have been confirmed in either natural or cultural systems. Second, attractors exist in phase space, not real space, so the identification of a strange attractor in real space is an additional step removed from understanding the social dynamics. Third, the identification and analysis of chaotic dynamical systems seem to require data sets that are larger and more precise than those commonly available to social scientists. Nevertheless, the application of chaos theory to the social sciences is in its infancy, and many surprises may await us.

CHAPTER 5. ADVANCED TOPICS
IN FRACTAL ANALYSIS

So far in this book, we have provided a basic introduction that presupposes little or no background on the subject. We have outlined and illustrated the central ideas that we see as essential to social scientists who wish to undertake fractal analysis of the data they study. We hope that Chapters 1 through 3 have provided the explanations that you need in order to embark on your own analyses.

Some years ago, Mandelbrot recognized a paradox inherent in the study of fractals. The essential ideas and elementary techniques are easy to understand but the underlying topology and advanced techniques are abstruse.

> The "core" of fractal geometry is simple and has become widely familiar, even to many middle-school students. But away from the core, the complication of the pictures and the difficulty of the problems increase sharply and suddenly. Very often, they jump up together, from very low to very high. (Mandelbrot, 2002, p. 10)

As a result, it can be easy to conduct a relatively naive analysis but sometimes hard to perform a thorough, precise, and convincing study. This same paradox, however, makes it challenging to write an introductory text that is detailed enough to be useful but at the same time simple enough to be comprehensible.

To achieve this, we therefore now wish to go beyond the simplest techniques and expand upon the basic ideas already introduced, to sketch a broader portrait of the variety of concepts and techniques available. The literature on fractals and their analysis is extensive and complex. Much of what's been written is more properly thought of as mathematics rather than statistics. A bewildering assortment of approaches has been devised, and myriad applications to a multiplicity of phenomena have been proposed. So, in this chapter, we will provide an overview of some of the beautiful flowers in this hothouse of techniques. We have selected several additional topics in fractal analysis to discuss: fractals embedded in three dimensions, self-affinity, fractal time series, multifractals, and lacunarity. On the whole, we present these topics conceptually and eschew some of the mathematics so as not to overwhelm the novice. Indeed, several of these topics are quite extensive themselves and could fill volumes of their own, so a full exegesis is impossible in the space available. Of course, the references cited develop the topics in greater detail.

Multiscaling Fractal Patterns

It sometimes happens that you discover a fractal pattern that displays more than one power law exponent such that there are different fractal dimensions governing different regions of the domain. We call these patterns "multiscaling" fractals. Figure 5.1 is a sketch of a graph illustrating a hypothetical situation such as this. Instead of a simple straight line, we see that the double-log plot presents two contiguous line segments with different slopes. Because the slope of the line is related to the fractal dimension, we can conclude that the fractal dimension changes at some critical value. This graph is telling us that the scaling relationship shifts from one regime to another, which implies that different processes or mechanisms are at work. Further investigation or observation may reveal why the fractal behavior of the object alters at that particular size or scale. Knowledge of the different scaling regimes can clearly contribute to understanding the behavior of the phenomenon and can have practical consequences, such as influencing the scales at which sampling is performed (Burrough, 1981).

A simple example is provided by Kennedy and Lin (1988) in their study of archaeological projectile point shapes. In archaeology, "projectile point" is the generic term used to denote stone arrowheads, spear points, dart points, and other similar armatures. They are manufactured by chipping away at an appropriate piece of rock, such as flint (chert), obsidian, quartz, and so on, until the desired form is achieved. The shapes of projectile points are important to archaeologists because they play a role in defining ancient cultures and periods. One might expect that the forms of projectiles would be tightly constrained by their function, yet they do vary quite a bit. Some of the variation is technological (such as changes in hafting methods), but some seems to be stylistic as well. The form of a projectile point is dominated overall by the shape of the blade (the pointy part that culminates in the tip) and the hafting element, or base. For example, the blade may be triangular or have concave or convex edges. A hafting element may be concave, flat, stemmed, or notched in various ways. On a smaller scale, the style of projectile points is often reflected in the way in which they were chipped. Some have long parallel chips removed that cross the whole blade. Others may have shorter chips removed that meet along the spine of the blade. When Kennedy and Lin performed their fractal analysis of projectile point forms, they found that the outlines of the edges of the projectile points, measured in plan view, exhibited two fractal dimensions, much as in Figure 5.1. The two different dimensions indicate that two different processes have caused the different slopes of the line. Kennedy and Lin's interpretation was that at larger scales, the fractal analysis described the

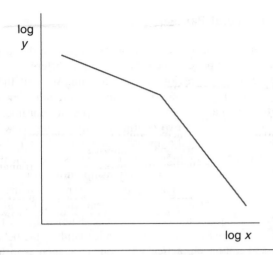

Figure 5.1 Graph of a hypothetical fractal relationship displaying two fractal dimensions over different parts of its domain.

scaling properties of the overall shape of the tool, whereas at smaller scales, the analysis captured the fractal behavior of the chipping pattern. In a more general sense, at larger scales, the fractal dimension quantified the overall "structure" of the pattern, whereas at smaller scales, the dimension quantified the "texture" of the pattern, and the structure and texture had different scaling properties.

There is no theoretical reason why an object or phenomenon cannot have more than two fractal dimensions operating at different scales, but obviously, as the number of distinct scaling regimes increases, the log-log plot will begin to resemble a curve. As their number multiplies, moreover, their meaningfulness diminishes rapidly. It is rarely possible in social science to observe a phenomenon over more than a few orders of magnitude, and we would prefer to observe fractal scaling behavior over at least one or two orders of magnitude in each case to be certain that it is truly exhibiting self-similarity and scale invariance. So in practical terms, the credible upper limit on different fractal dimensions for different scaling regimes is probably 3 or 4. Moreover, in any particular case, as the number of fractal dimensions increases, their interpretation may become more obscure. That is, it may become difficult to plausibly explain the existence of several different scaling regimes. If that explanation becomes complicated and implausible, then the fractal model starts to lose its appeal as parsimonious and elegant.

To avoid confusion, please note that multiscaling is not the same as multifractals, which are discussed later. In addition, when we say "different fractal dimensions," we mean different *values* of the Hausdorff dimension, not other kinds of fractal dimensions (such as the information dimension or correlation dimension), which also exist and are involved in the analysis of multifractals.

Patterns Embedded in Three Dimensions

Fractal analysis in the social sciences is not limited to patterns embedded in one or two dimensions. Fractals embedded in three dimensions also exist. The classic example of this type of pattern is the rough surface. Surface roughness has been investigated intensively using fractal analysis. Indeed, the American Society of Mechanical Engineers has promulgated a fractal method for quantifying surface roughness (American Society of Mechanical Engineers, 2002). The most famous example of fractal surface roughness is geographic relief and topography (Mandelbrot, 1975; 1983, pp. 256–276), which led Mandelbrot and Richard Voss to create their famous simulations of landscapes and planets.

Although different kinds of fractals embedded in three dimensions are known, some of these may be visualized as rough surfaces. Fracture surfaces, for instance, typically exhibit fractal roughness. In social science, several such phenomena have been studied. For example, the densities of various social measures can be analyzed in the same fashion as topographic surfaces. The social variable is treated as an elevation or altitude on the z-axis. Batty and Longley (1994) have, in this way, examined phenomena such as population densities. Wong and his colleagues (Wong, Lasus, & Folk, 1999) have examined how racial segregation varies spatially with respect to scale and found fractal patterns there as well.

As usual, various methods are available to analyze these kinds of fractal patterns. We already mentioned in Chapter 3 that box counting can be extended to three dimensions to analyze these kinds of patterns, but other methods exist as well. Methods based on the variograms or the power spectra of surfaces are particularly common in geology and geography (Burrough, 1981; Klinkenberg, 1994; Wen & Sinding-Larsen, 1997). As usual, the method to be preferred may depend upon issues such as the form of the available data (e.g., whether the data points are sampled at regular or irregular intervals), the type of fractal pattern being investigated (e.g., a continuous crumpled surface versus a collection of discontinuous objects in a volume), and so forth.

Self-Affine Fractals

The issue of fractals embedded in three dimensions brings us naturally to the topic of self-affinity. If all fractals were exactly and perfectly self-similar, the

68

world would look immensely complicated but still rather boring and predictable. We have already seen that most real-world fractals exhibit their self-similarity statistically rather than strictly. Some phenomena exhibit their fractal behavior in yet more complex ways. There are, in fact, a number of other ways in which fractal patterns can occur.

Sometimes, a phenomenon will exhibit one type of fractal behavior in one direction and a different type of fractal behavior in another. Topographic relief, for example, scales by different factors in different directions. In the two horizontal directions, landscape is isotropic and statistically self-similar. This is reflected in coastlines and contour lines, which, as we have seen, tend to have characteristic fractal dimensions around 1.26. Elevation or altitude is also fractal, but it scales by a much smaller characteristic scaling factor than the horizontal component. If you have ever seen a topographic map that includes a vertical section or profile of the landscape, the latter probably has some vertical exaggeration so that the relief is clearer. The vertical exaggeration is a reflection of the disjuncture between the natural horizontal and vertical scaling factors. When different aspects of a single phenomenon scale differently like this, it is said to exhibit "self-affinity" rather than true self-similarity.

Fractal Time Series

The analysis of time series is a distinct—and expansive—topic in statistics, and the situation is no different in the field of fractal statistics. We can touch upon only a couple of basic concepts here, which does not really do justice to the subject; it really merits a volume of its own. Fractal time series are generally self-affine, so we have chosen to describe them after introducing the concept of self-affinity.

A typical time series is a single variable y measured at fixed intervals of time t. There are special techniques for analyzing time series when the time intervals are irregular rather than fixed, but we will not consider them here. Time series are usually graphed with t on the horizontal axis and y on the vertical axis. If y is a random variable whose values are drawn from a standard normal distribution, then the signal is random "white noise." In this case, the movements in time will be independent and therefore uncorrelated, and the average deviation from the mean will be zero because positive and negative shifts will cancel out.

One way to think about such a time series is as a model of Brownian motion. The term *Brownian motion* refers to the random motions of particles suspended in a colloid, and it is named for botanist Robert Brown. Brown (1828) observed pollen suspended in water under a microscope and saw that the particles jiggled irregularly. The motion seemed to imply that pollen was alive. Brown also ground many materials, including many kinds of stone, to a fine powder, and observed them under the same microscope. Again, he could see minute particles moving erratically, although the stone was not living matter. So he understood that he had observed a physical process, but he falsely believed that he had seen the motion of the constituent molecules of these materials.

Albert Einstein developed our modern understanding of Brownian motion in statistical mechanics. His immense fame rests so largely on his theories of relativity that we commonly forget that one of his groundbreaking 1905 papers was on Brownian motion (Einstein, 1956). In this early paper, he infers that the visible Brownian motion of the small particles is caused by collisions with molecules or atoms, thus providing support for the atomic theory of matter and demonstrating fundamental relationships with heat and diffusion. He also shows how these motions obey normal, Gaussian statistics (Einstein, 1956, p. 16).

Today, Brownian motion supplies a basic model for the study of time series. The running sum in time of our y values can be thought of as displacements representing Brownian motion. The arithmetic mean of the displacement will be zero, but the root mean square displacement is proportional to the number of steps. So, the movement in our y variable is related to the linear progress of time t by a factor proportional to the square root of y. If we wish to rescale the graph of our time series, we will have to take into account this relationship between time and displacement. If we scale time by a factor of, say, 2, then we will have to rescale the y variable, the amplitude (i.e., height on the y-axis), by $\sqrt{2}$. More generally, if we scale time t by a factor r, then we will have to rescale the amplitude y by $\sqrt{r}$ or equivalently $r^{1/2}$ (Peitgen et al., 1992, pp. 481–493). This scaling exponent is usually called the "Hurst exponent" and is symbolized by the letter H. Note that the existence of different scaling factors in the horizontal and vertical dimensions tells us that this is a self-affine process.

Classical Brownian motion turns out to be a special, indeed, a unique case of a more general phenomenon called *fractional Brownian motion*. The scaling properties of classical Brownian motion in a time series are described by $H = \frac{1}{2}$, but other values of H are possible. H can take any value between 0 and 1, but three different regimes exist. The first regime is for $H = \frac{1}{2}$, ordinary Brownian motion. For $H < \frac{1}{2}$ the changes in y are negatively correlated, so there is an increased probability that each succeeding value of the series will be different from its predecessor. So, low values will tend to follow high ones, and vice versa. This is often called *anti-persistence* because the tendency is for the trace to reverse direction. This leads to highly erratic fluctuations in the time series, and the graph of it looks very rough—rougher than ordinary Brownian motion. For $H > \frac{1}{2}$ the changes in

...*y* are positively correlated, and therefore, the tendency is for values to be succeeded by similar values more often than chance alone would dictate (Figure 5.2). So, high values tend to be followed by more high values, whereas low values tend to be followed by more low values. This behavior is often called *persistence,* and it leads to smoother curves than one obtains from classical Brownian motion. In both cases, $H < \frac{1}{2}$ and $H > \frac{1}{2}$, the time series is said to exhibit long-term memory because the variation in the increments includes autocorrelations; they are not independent and uncorrelated as they are in ordinary Brownian motion. The Hurst exponent is related to the fractal dimension by a simple relation: $D = 2 - H$. When $H \ne \frac{1}{2}$, the time series represents fractional Brownian motion and has fractal, not Gaussian, properties. Such time series cannot be analyzed using Gaussian statistics. Therefore, we study how their variance depends upon the time interval over which it is measured (Liebovitch & Yang, 1997, p. 4557).

Fractional Brownian motion derives its importance from the fact that many processes have traditionally been modeled using Brownian motion or some closely akin process based on the normal curve and parametric statistics. Harold E. Hurst, who was a hydrologist, developed the exponent now named after him in the 1950s while conducting studies on the flow of the Nile River. He discovered empirically that the time series of many natural

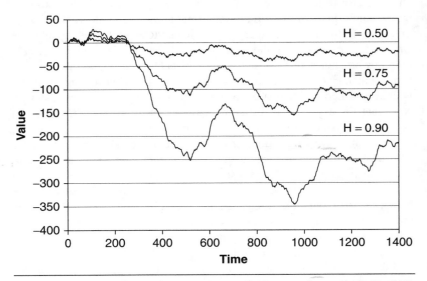

Figure 5.2 Graph of time series for $H = 0.5$ (classic Brownian motion), $H = 0.75$, and $H = 0.9$. For $H > 0.5$, one can see that the curve is smoother than that of classical Brownian motion, an indication of "persistence," a form of long-term memory or autocorrelation.

phenomena exhibited scaling exponents in the .7 to .8 range, thus clearly diverging from the independent and uncorrelated expectations of a random model. In the 1960s, Mandelbrot and Wallis (1968) elaborated on Hurst's empirical findings and generalized the ideas to random Gaussian noises and fractional Brownian motion (Turcotte, 1997, p. 159). Since then, these formulations have found wide applicability in the natural and social sciences. In the social sciences, this approach has been used most frequently to study financial time series, such as stock, commodity, and currency prices. Mandelbrot (1963a, 1963b) pioneered the study of scaling in finance with historical studies of cotton prices in the early 1960s. He has continued to investigate financial time series and has written prolifically about them (Mandelbrot, 1997, 2002, 2005). SEE THESE PAPERS !!!

Fractal statistics have become popular in finance and have even helped spawn a new field called "econophysics" (Bouchard & Potters, 2003; Mandelbrot, 2005; Mantegna & Stanley, 2000; Voit, 2003). Its popularity can be explained in part by the fact that fractal statistics often model the data better than conventional Gaussian statistics. Fractal models of finance have weighty implications because they undercut various cherished assumptions of classical econometric models such as the efficient market hypothesis. The literature on fractals in finance and economics is much too large to review here, but the references cited earlier provide a starting point for the interested reader. Fractal time series have been studied using the Hurst exponent in other domains of social science as well. Verma (1998), for example, studied the timing of emergency calls to the police using the Hurst exponent and found that the time series was fractal with exponents $H > \frac{1}{2}$, indicating a persistent, non-normal process.

Several methods exist for estimating the Hurst exponent for empirical time series. The oldest and best known method is "rescaled range analysis," symbolized R/S. Here's how to calculate the Hurst exponent using R/S analysis. Let's say that we have a time series with $N = 128$ observations of a particular variable y. The running sum of the deviations from the mean is

$$y_N = \sum_{i=1}^{N} (y_i - \bar{y}_N), \qquad (5.1)$$

where $\bar{y}$ is the mean. The range of this running sum of the deviations is

$$R_N = (y_N)_{max} - (y_N)_{min} \qquad (5.2)$$

and the standard deviation is

$$S_N = \left[\frac{1}{N} \sum_{i=1}^{N} (y_i - \bar{y})^2 \right]^{\frac{1}{2}}, \qquad (5.3)$$

Shows scaling properties of fractal time series; a "H" !!!

72

where $\bar{y}$ is the mean. Then the Hurst exponent is

$$(R_N/S_N) = (N/2)^H. \tag{5.4}$$

We estimate this relation by calculating R_N and S_N for progressively smaller, adjacent, nonoverlapping subsets of the series. Start by calculating R_N and S_N for the whole series of 128 observations. Then subdivide the series into two subseries of 64 observations ($i = 1, 2, \ldots 64$, and $i = 65, 66, \ldots 128$). Calculate R_{64} and S_{64} for each of the two subseries and divide to find (R_{64}/S_{64}) for each subseries, and finally average the two to obtain (R_{64}/S_{64})$_{av}$. Next, subdivide the original set of observations into four subseries of 32 observations each and repeat the process to obtain four values of (R_{32}/S_{32}). Average those four values to obtain (R_{32}/S_{32})$_{av}$. Repeat this process for $n = 16, 8,$ and 4. Plot the values of the logarithm of (R_N/S_N)$_{av}$ against the logarithm of $N/2$ (i.e., $64, 32, 16 \ldots$). The slope of the best-fit line provides an estimate of H (Turcotte, 1997, pp. 158–161).

Many other methods have been developed for studying the scaling properties of fractal time series. For example, the variogram method can be used with time series as well as with surfaces. The variogram is a common geostatistic. It evaluates how the average squared distances between the amplitudes of points vary in relation to the horizontal distance between the points. In a time series, this measure obviously relates to how the variance changes along the time axis, so the analogy to the Hurst exponent is apparent. Other methods include detrended fluctuation analysis (Peng et al., 1994), a roughness-length method (Malinverno, 1990), a variance analysis method, and a zero-crossing analysis method (Liebovitch & Yang, 1997). Various methods have been compared and evaluated by Liebovitch and Yang (1997), Liebovitch, Penzel, and Kantelhardt (2002), Bassingthwaighte and Raymond (1994), and Klinkenberg (1994). A discussion of the merits of various methods is beyond the scope of this text, but it should be noted that the rescaled range method is sensitive to the length of the time series data available for analysis; long series are strongly preferred for accurate estimation.

Multifractals

Early in this volume, we defined fractals as sets that possess certain important characteristics, namely, self-similarity and fractional dimension. Multifractals represent the extension of fractal theory from sets to *measures*. Membership in a set has a dichotomous or Boolean quality: A thing is either an element of a set or it is not. In contrast to a set, a *measure* incorporates the idea of quantity or magnitude. Therefore, *measure* is a more general mathematical concept than *set.* Multifractals are the application of the principle of self-similarity and the attendant apparatus of fractal concepts to measures (Evertsz & Mandelbrot, 1992).

So, multifractals address not presence/absence questions, but rather questions of quantity, density, concentration, and so forth. Therefore, they are appropriate in situations when a thing is present everywhere but in varying amounts; in such a case, the regular fractal dimension is not a helpful characterization because all the boxes in the covering will be filled. Multifractals, by using a measure, take into account the quantity present.

The (mono)fractals we have discussed through this book are characterized by a single parameter, the fractal dimension. Multifractals are characterized by a spectrum of related dimensions. The lowest order dimension is the fractal dimension, and the higher order dimensions are essentially the higher order moments of the fractal dimension. If the different dimensions are all the same, then the phenomenon is called a *homogenous fractal.* If the dimensions vary but are well-defined power laws, then the phenomenon classifies as a *multifractal.* If the dimensions are not power laws, then phenomenon is not multifractal at all (Turcotte, 1997, pp. 113–129).

The multifractal approach has been applied to a variety of natural phenomena (see references in Turcotte, 1997, p. 129), but it seems to have found few adherents as yet among social scientists. Perhaps the mathematics seem formidable, or maybe the techniques have simply not diffused from their source in mathematics and physics over into social science territory. As a practical matter, multifractal analysis is not *exceedingly* more difficult than monofractal analysis, so practitioners should not be too discouraged from attempting it. On the other hand, multifractal analysis does present certain challenges or disadvantages. One of the most attractive characteristics of monofractal analysis is its simplicity and parsimony. It is undeniably elegant. But where a monofractal distribution is characterized by perhaps two constants (see Equation 1.2), a full spectrum of multifractal dimensions produces in theory an infinite number of constants (Turcotte, 1997, p. 129). Clearly, much of the attractive parsimony is lost with multifractals. In addition, multifractals are not necessarily scale invariant and, indeed, are unlikely to be so (Turcotte, 1997, p. 128). Turcotte (1997) has concluded that "multifractal statistics are much less useful than monofractal statistics from a practical point of view" (p. 129).

Lacunarity

Fractals with the same fractal dimension can look radically different. Thus, the fractal dimension, although it is the invariant parameter for fractal relations, does not completely characterize a particular fractal pattern, any more than the mean uniquely or completely characterizes a particular sample from a normal distribution. So, in some cases, portraying a fractal solely in terms of its dimension may seem to be a less than adequate specification because it fails to disambiguate obviously different patterns. To more fully describe fractal patterns, Mandelbrot (1983, pp. 310–318; Gefen, Aharony, & Mandelbrot, 1984) proposed a measure called *lacunarity* to express the texture of fractals. Lacunarity specifies the distribution of gaps in the patterns. By describing their "clumpiness," lacunarity offers a fuller statistical portrayal. A number of approaches to measuring lacunarity have been developed (Lin & Yang, 1986; Turcotte, 1997, pp. 109–112), but the sliding window technique developed by Allain and Cloitre (1991) is the most widely used because it is relatively easy to understand and implement. Lacunarity is actually a nonfractal, scale-dependent technique, so it can be used to quantify any pattern, not just fractal ones. A detailed discussion of lacunarity is beyond the scope of this monograph, so we will note only that in combination with the fractal dimension, lacunarity can be used to describe fractal patterns further (Brown et al., 2005, pp. 57–58).

Conclusion

As we have seen in this chapter, fractal patterns of many different types can be defined and measured using diverse techniques. A variety of statistics can be used to study fractal data, including the mean, variance, standard deviation, relative dispersion (standard deviation/variance), Fano factor (variance/mean), mean squared deviation, and rescaled range. The lesson is that the

> statistical properties, such as the average or variance, of a fractal object or process depend on the resolution used to measure them. Thus *it does little good to measure these statistical properties at only one resolution.* The meaningful measurement is *to determine how these statistical properties depend on the resolution used to measure them.* (Liebovitch, 1998, p. 98, emphasis in original)

In other words, because statistics such as mean and variance are not consistent or stable estimators of fractal characteristics, in order to identify stable estimators, we must examine how our statistics vary in relation to the scale of observation or measurement. The techniques described in this chapter do precisely that. Of course, we have only scratched the surface of the literature on these more complex topics in fractal analysis. Many other techniques have been developed for different kinds of phenomena in varying situations. We only hope that we have opened a small window onto this large field and that the view will encourage our colleagues to explore the territory.

CHAPTER 6. FINAL CONSIDERATIONS

We believe that many significant social phenomena are fractal. Fractal patterns are seemingly common because roughness, in all of its manifestations, is ubiquitous. Fractal analysis measures roughness directly and parsimoniously (Mandelbrot, 2005, p. 194). This is contrary to the traditional approach in science, which is to see complexity and attribute it to randomness, to try to smooth it, and to try to model nonlinearity with successively more complicated approximations to linearity. When we do these things, we idealize away the substance of what we are studying. It would be irresponsible to ignore statistical approaches, typified by fractal statistics, that measure roughness directly. But of course, this doesn't mean that everything is fractal. Many social phenomena produce Gaussian statistics, exponential relations, or other nonfractal patterns. Therefore, each case must be argued and evaluated on its own merits.

> Most emphatically, I do not consider the fractal point of view as a panacea, and each case analysis should be assessed by the criteria holding in its field, that is, mostly upon the basis of its powers of organization, explanation, and prediction, and not as example of a mathematical structure. (Mandelbrot, 1983, p. 3)

Thus, "Mandelbrot, like Prime Minister Churchill before him, promised not Utopia, but blood, sweat, toil, and tears" (Cootner, 1964, cited in Mandelbrot, 2002, p. 14).

A fair evaluation on the merits, however, does not start from the presumption that data should be normal (or Poisson or binomial, etc.), even if that has been the traditional interpretation. There are no grounds for assuming that all distributions should be normal. Many *are* normal, but only because normality is the inevitable consequence of certain stochastic processes. By the same token, power laws are the inevitable consequence of many simple processes, including those described in Chapter 4. In fact, fractal statistics provide fertile ground for null hypotheses, although unfortunately, significance tests are still largely lacking (but see Jin & Frechette, 2004).

Should I Try Fractal Analysis?

So, if fractal patterns are common but by no means universal, then how do you know when to attempt a fractal analysis? Because fractals are, generally speaking, complicated patterns, when we look at a data set, we first ask ourselves, "Are there elements or pieces or fluctuations of many different sizes or magnitudes, and if so, are there lots of little ones and a few big ones?" or, alternatively, "Are there highly irregular shapes involved?" If the answer to one of these questions is "yes," then it is probably worth examining whether the phenomenon has fractal characteristics. If it does, then the fractal dimension is the proper statistic to describe the pattern.

Another practical approach when studying a complex social phenomenon is to start conceptually and ask yourself whether it has self-similar properties. Does the big thing (labor union, corporation, nonprofit, neighborhood, sodality, clan, government, language family, etc.) have smaller parts similar in structure to the whole? Or, is there a process that generates similar elements of widely different or ever declining magnitude? If the answer is "yes," then you can consider what data (or what statistics derived from the data) could reflect the fractal nature of the phenomenon.

Moving beyond the descriptive to the explanatory, remember that although fractals are sets, and the fractal dimension is a descriptive statistic, fractals are integrally linked to dynamics and therefore provide clues about the dynamical processes that produce them. Nevertheless, when you discover a fractal phenomenon, it is wise to ask yourself what the fractal characteristics convey about the phenomenon. Does it tell you something new about its structure or about the dynamics that produced it?

Of course, one should not underestimate the descriptive or exploratory role of fractal analysis. It is common enough to hear fractals disparaged as "only" descriptive, not explanatory. This attitude unfairly minimizes the importance of proper description. It is all very well to insist on explanation, but description normally comes first, both historically and logically. Developing an accurate description of the phenomenon you are trying to understand is indispensable. How likely is one to arrive at a correct explanation if one can't describe the data correctly? In many cases, only fractal statistics provide an accurate description. It is neither fair nor reasonable to expect, much less insist upon, an a priori theoretical justification for considering them, any more than one would expect a complex theoretical justification for the use of traditional parametric statistics.

Caveats

We should not understate the obstacles that you may encounter in conducting fractal analysis. Fractal analysis is still fairly new in some social science fields, and you may meet resistance in attempting innovative applications of it. It is good to be aware of some of the pitfalls and criticisms that have been leveled at fractal analysis.

It is sometimes difficult to demonstrate power law behavior in data sets because fitting data to models can be challenging. In Chapter 2, although we discussed some of the complexities of fitting and validating power law distributions as models of empirical data, we did not exhaust the topic. Additional methods also exist—maximum likelihood estimation is only one (Clauset et al., 2007)—and it may be necessary to use them. More fundamentally, it can be difficult to ascertain which of several distributions provides the best fit to a data set. For example, lognormal distributions can be especially tricky to differentiate from power laws because of their innate similarities (Mitzenmacher, 2003). Stretched exponentials are also difficult to distinguish from power laws. Part of the problem is that these kinds of distributions differ mainly in the shapes of their tails, which is where the fewest data points lie. Because the tail encompasses the rarest and most extreme events, it is difficult to define precisely and is subject to the greatest errors because of the small numbers involved. In short, the problem of establishing with certainty whether a power law relation exists can be complicated, and it may even be impossible to resolve completely. Thus, one ought to take into account the power of the statistics used to support one's inferences because some are weaker than they appear. For example, Durlauf (2005, pp. F232–F233) has pointed out one weakness in applying Zipf's rank-size rule by regressing rank against size. Because rank is constructed to decrease inversely with size, the finding of a negative exponent is relatively uninformative; neither do we know as much as we would like about how the results of this kind of analysis would vary given different, alternative processes. Another weakness in some fractal analyses arises when the authors propose a fractal model but neglect to test plausible alternative models. It is advisable to preclude criticism by showing not only that a fractal model fits but that it fits better than other models that might reasonably be applied to the data.

Turning from the mathematical to the social, one criticism that has been leveled at some fractal models in the social sciences is that they have been predicated upon weak behavioral assumptions or implausible social mechanisms (Durlauf, 2005). Social scientists themselves are much less likely to fall into this trap than natural or physical scientists who sometimes attempt to transfer their models to social data sets without taking into account the unique characteristics of human action.

We encourage you to look for fractal patterns in the data sets with which you're familiar. When you find fractals, we urge you to explore them using the kinds of techniques outlined in this monograph. Much of social life and cultural behavior is fractal, which helps make it varied and interesting. Fractal analysis alone will not unlock all the secrets of society and culture, but at a minimum, recognizing fractal patterns constitutes an important advance in understanding the complex patterns of human behavior.

There are many possible directions for future applications of fractal analysis in the social sciences. As we mentioned in the preface, a systematic review of fractal social phenomena seems overdue. Research on fractals seems quite active in economics and geography, but less so in sociology, anthropology, and political science. We hope that this short text assists, or even inspires, some researcher to explore the many fractal patterns embedded in the human experience.

REFERENCES

Abbott, A. (2001). *Chaos of disciplines.* Chicago: University of Chicago Press.

Allain, C., & Cloitre, M. (1991). Characterizing the lacunarity of random and deterministic fractal sets. *Physical Review A, 44*(6), 3552–3558.

American Society of Mechanical Engineers. (2002). *Surface texture (surface roughness, waviness, and lay): An American national standard.* New York: Author.

Arlinghaus, S. L. (1985). Fractals take a central place. *Geografiska Annaler, 67*B(2), 83–88.

Arlinghaus, S. L. (1993). Central place fractals: Theoretical geography in an urban setting. In N. Siu-Ngan Lam & L. DeCola (Eds.), *Fractals in geography* (pp. 213–227). Engelwood Cliffs, NJ: Prentice Hall.

Axtell, R. L. (2001). Zipf distribution of firm sizes. *Science, 293,* 1818–1820.

Bak, P. (1997). *How nature works: The science of self-organized criticality.* New York: Oxford University Press.

Barabási, A.-L. (2002). *Linked: The new science of networks.* Cambridge, MA: Perseus.

Barabási, A.-L. (2005). The origin of bursts and heavy tails in human dynamics. *Nature, 435,* 207–211.

Barabási, A.-L., & Albert, R. (1999). Emergence of scaling in random networks. *Science, 286,* 509–512.

Barton, C., & Nishenko, S. (2003). Natural disasters—Forecasting economic and life losses [Online]. Available at: http://pubs.usgs.gov/fs/natural-disasters

Bassingthwaighte, J. B., Liebovitch, L. S., & West, B. J. (1994). *Fractal physiology.* New York: Oxford University Press.

Bassingthwaighte, J. B., & Raymond, G. M. (1994). Evaluating rescaled range analysis for time series. *Annals of Biomedical Engineering, 22,* 432–444.

Batty, M. (2008). The size, scale, and shape of cities. *Science, 319,* 769–771.

Batty, M., & Longley, P. (1994). *Fractal cities: A geometry of form and function.* New York: Academic Press.

Benguigui, L. (1995). A fractal analysis of the public transportation system of Paris. *Environment and Planning A, 27,* 1147–1161.

Benguigui, L., Czamanski, D., Marinov, M., & Portugali, Y. (2000). When and where is a city fractal? *Environment and Planning B: Planning and Design, 27,* 507–519.

Benguigui, L., & Daoud, M. (1991). Is the suburban railway system a fractal? *Geographical Analysis, 23,* 362–368.

Berry, B. J. L. (1967). *Geography of market centers and retail distribution.* Englewood Cliffs, NJ: Prentice Hall.

Block, A., von Bloh, W., & Schellnhuber, H. J. (1990). Efficient box-counting determination of generalized fractal dimensions. *Physical Review A, 42,* 1869–1874.

Bouchard, J.-P. (2008). Economics needs a scientific revolution. *Nature, 455*(30), 1181.

Bouchard, J.-P., & Potters, M. (2003). *Theory of financial risk and derivative pricing: From statistical physics to risk management* (2nd ed.). Cambridge, UK: Cambridge University Press.

Bren, L. J. (1995). Aspects of the geometry of riparian buffer strips and its significance to forestry operations. *Forestry Ecology and Management, 75,* 1–10.

Brown, C. T. (2001). The fractal dimensions of lithic reduction. *Journal of Archaeological Science, 28*(6), 619–631.

Brown, C. T., Liebovitch, L. S., & Glendon, R. (2007). Lévy flights in Dobe Ju/'hoansi foraging patterns. *Human Ecology, 35*(1), 129–138.

Brown, C. T., Witschey, W. R. T., & Liebovitch, L. S. (2005). The broken past: Fractals in archaeology. *Journal of Archaeological Method and Theory, 12*(1), 37–78.

Brown, R. (1828). Mr. Rob. Brown's brief account of microscopial observations made in the months of June, July, and August, 1827, on the particles contained in the pollen of plants; and on the general existence of active molecules in organic and inorganic bodies. *The Philosophical Magazine, or Annals of Chemistry, Mathematics, Astronomy, Natural History, and General Science, 4*(21), 161–173.

Buczkowski, S., Kyriacos, S., Nekka, F., & Cartilier, F. (1998). The modified box-counting method: Analysis of some characteristic parameters. *Pattern Recognition, 31*, 411–418.

Burnett, P., & Pongou, R. (2006). Network underpinnings of behavioral travel demand: Fractal analysis of Boston's transportation system. *Transportation Research Record: Journal of the Transportation Research Board, 1985*, 241–247.

Burrough, P. A. (1981). Fractal dimensions of landscapes and other environmental data. *Nature, 294*, 240–242.

Cederman, L.-E. (2003). Modeling the size of wars: From billiard balls to sand piles. *American Political Science Review, 97*(1), 135–150.

Christaller, W. (1966). *Central places in southern Germany*. Englewood Cliffs, NJ: Prentice Hall. (Original work published 1933)

Clauset, A., Shalizi, C. R., & Newman, M. E. J. (2007). Power-law distributions in empirical data. arXiv:0706.1062v1 [physics.data-an].

Clauset, A., Young, M., & Gleditsch, K. S. (2007). On the frequency of severe terrorist events. *Journal of Conflict Resolution, 51*, 58–87.

Crawford, T. W., Commito, J. A., & Borowik, A. M. (2006). Fractal characterization of *Mytilus edulis L.* spatial structure in intertidal landscapes using GIS methods. *Landscape Ecology, 21*, 1033–1044.

De Cola, L., & Lam, N. S.-N. (1993). Introduction to fractals in geography. In N. S.-N. Lam & L. De Cola (Eds.), *Fractals in geography* (pp. 3–22). Englewood Cliffs, NJ: Prentice Hall.

Durlauf, S. N. (2005). Complexity and empirical economics. *Economic Journal, 115*, F225–F243.

Edwards, A. M., Phillips, R. A., Watkins, N. W., Freeman, M. P., Murphy, E. J., Afanasyev, V., Buldyrev, S. V., da Luz, M. G. E., Raposo, E. P., Stanley, H. E., & Viswanathan, G. M. (2007). Revisiting Levy flight search patterns of wandering albatrosses, bumblebees and deer. *Nature, 451*, 1044–1049.

Einstein, A. (1956). *Investigations on the theory of the Brownian movement* (A. D. Cowper, Trans.). New York: Dover.

Englehardt, J. D. (2002). Scale invariance of incident size distributions in response to sizes of their causes. *Risk Analysis, 22*(2), 369–381.

Evertsz, C. J. G., & Mandelbrot, B. B. (1992). Appendix B: Multifractal measures. In H.-O. Peitgen, H. Jürgens, & D. Saupe, *Chaos and fractals: New frontiers of science* (pp. 921–953). New York: Springer-Verlag.

Falconer, K. J. (2003). *Fractal geometry: Mathematical foundations and applications* (2nd ed.). Chichester, UK: Wiley.

Foroutan-pour, K., Dutilleul, P., & Smith, D. L. (1999). Advances in the implementation of the box-counting method of fractal dimension estimation. *Applied Mathematics and Computation, 105*, 195–210.

Fujita, M., Krugman, P., & Venables, A. (1999). *The spatial economy: Cities, regions, and international trade*. Cambridge, MA: MIT Press.

Gefen, Y., Aharony, A., & Mandelbrot, B. B. (1984). Phase transitions on fractals: III. Infinitely ramified lattices. *Journal of Physics A, 17*, 1277–1289.

Gomes, M. A. F., Vasconcelos, G. L., Tsang, I. J., & Tsang, I. R. (1999). Scaling relations for diversity of languages. *Physica A, 271*, 489–495.

Goodchild, M. F. (2004). The validity and usefulness of laws in geographic information science and geography. *Annals of the Association of American Geographers, 94,* 300–303.

Goodchild, M. F., & Mark, D. M. (1987). The fractal nature of geographic phenomena. *Annals of the Association of American Geographers, 77,* 265–278.

Hamilton, M. J., Milne, B. T., Walker, R. S., Burger, O., & Brown, J. H. (2007). The complex structure of hunter-gatherer social networks. *Proceedings of the Royal Society B, 274,* 2195–2202.

Hastings, H. M., & Sugihara, G. (1993). *Fractals: A user's guide for the natural sciences.* New York: Oxford University Press.

Jeong, H., Neda, Z., & Barabási, A.-L. (2003). Measuring preferential attachment for evolving networks. *Europhysics Letters, 61,* 567–572.

Jin, H. J., & Frechette, D. L. (2004). A new *t*-test for the R/S analysis and long term memory in agricultural commodity prices. *Applied Economics Letters, 11,* 661–667.

Kennedy, S. K., & Lin, W.-H. (1988). A fractal technique for the classification of projectile point shapes. *Geoarchaeology, 3*(4), 297–301.

Klinkenberg, B. (1994). A review of methods used to determine the fractal dimension of linear features. *Mathematical Geology, 26*(1), 23–46.

Kruger, A. (1996). Implementation of a fast box-counting algorithm. *Computer Physics Communications, 98,* 224–234.

Krugman, P. (1996). *The self-organizing economy.* Cambridge, MA: Blackwell.

Lam, N. S.-N., & De Cola, L. (1993). Fractal measurement. In N. S.-N. Lam & L. De Cola (Eds.), *Fractals in geography* (pp. 23–55). Englewood Cliffs, NJ: Prentice Hall.

Liebovitch, L. S. (1998). *Fractals and chaos simplified for the life sciences.* New York: Oxford University Press.

Liebovitch, L. S., Penzel, T., & Kantelhardt, J. W. (2002). Physiological relevance of scaling of heart phenomena. In A. Bunde, J. Kropp, & H. J. Schellnhuber (Eds.), *The science of disasters: Climate disruptions, heart attacks, and market crashes* (pp. 259–281). Berlin: Springer.

Liebovitch, L. S., & Scheurle, D. (2000). Two lessons from fractals and chaos. *Complexity, 5*(4), 34–43.

Liebovitch, L. S., Scheurle, D., Rusek, M., & Zochowski, M. (2001). Fractal methods to analyze ion channel kinetics. *Methods, 24,* 359–375.

Liebovitch, L. S., & Shehadeh, L. A. (2003). *The mathematics and science of fractals and chaos* [CD-ROM]. Boynton Beach, FL: Deco Bytes Education.

Liebovitch, L. S., & Todorov, A. T. (1996). Invited editorial on "Fractal dynamics of human gait: Stability of long-range correlations in stride fluctuations." *Journal of Applied Physiology, 80,* 1446–1447.

Liebovitch, L. S., Todorov, A. T., Zochowski, M., Scheurle, D., Colgin, L., Wood, M. A., Ellenbogen, K. A., Herre, J. M., & Bernstein, R. C. (1999). Nonlinear properties of cardiac rhythm abnormalities. *Physical Review E, 59,* 3312–3319.

Liebovitch, L. S., & Tóth, T. (1989). A fast algorithm to determine fractal dimensions by box counting. *Physics Letters A, 141*(8, 9), 386–390.

Liebovitch, L. S., & Tóth, T. (1990). The Akaike information criterion (AIC) is not a sufficient condition to determine the number of ion channel states from single channel recordings. *Synapse, 5,* 134–138.

Liebovitch, L. S., & Yang, W. (1997). Transition from persistent to antipersistent correlation in biological systems. *Physical Review E, 56*(4), 4557–4566.

Lin, B., & Yang, Z. R. (1986). A suggested lacunarity expression for Sierpinski carpets. *Journal of Physics A, 19,* L49–L52.

Lösch, A. (1954). *The economics of location* (Translated from the 2nd rev. ed. by W. H. Woglom with the assistance of W. F. Stolper). New Haven, CT: Yale University Press.

Lu, Y., & Tang, J. (2004). Fractal dimension of a transportation network and its relationship with urban growth: A study of the Dallas-Fort Worth area. *Environment and Planning B: Planning and Design, 31,* 895–911.

Malinverno, A. (1990). A simple method to estimate the fractal dimension of a self-affine series. *Geophysical Research Letters, 17*(11), 1953–1956.

Malmgren, R. D., Stouffer, D. B., Motter, A. E., & Amaral, L. A. N. (2008). A Poissonian explanation for heavy tails in e-mail communication. *Proceedings of the National Academy of Sciences, 105,* 18153–18158.

Mandelbrot, B. B. (1963a). New methods in statistical economics. *Journal of Political Economy, 71*(5), 421–440.

Mandelbrot, B. B. (1963b). The variation of certain speculative prices. *Journal of Business, 36,* 394–419.

Mandelbrot, B. B. (1967). How long is the coast of Britain? Statistical self-similarity and fractional dimension. *Science, 156,* 636–638.

Mandelbrot, B. B. (1975). Stochastic models for the Earth's relief, the shape and the fractal dimension of the coastlines, and the number-area rule for islands. *Proceedings of the National Academy of Sciences, 72,* 3825–3828.

Mandelbrot, B. B. (1977). *Fractals: Form, chance, and dimension.* San Francisco: W. H. Freeman.

Mandelbrot, B. B. (1983). *The fractal geometry of nature* (Rev. ed.). New York: W. H. Freeman.

Mandelbrot, B. B. (1997). *Fractals and scaling in finance.* Berlin: Springer.

Mandelbrot, B. B. (2002). *Gaussian self-affinity and fractals.* Berlin: Springer.

Mandelbrot, B. B. (2003). Multifractal power law distributions: Negative and critical dimensions and other "anomalies," explained by a simple example. *Journal of Statistical Physics, 111,* 739–774.

Mandelbrot, B. B. (2005). The inescapable need for fractal tools in finance. *Annals of Finance, 1,* 193–195.

Mandelbrot, B. B., & Wallis, J. R. (1968). Noah, Joseph, and operational hydrology. *Water Resources Research, 4,* 909–918.

Mantegna, R. N., & Stanley, H. E. (2000). *An introduction to econophysics: Correlations and complexity in finance.* New York: Cambridge University Press.

Meakin, P. (1986). Computer simulation of growth and aggregation processes. In H. E. Stanley & N. Ostrowsky (Eds.), *On growth and form: Fractal and non-fractal patterns in physics* (pp. 111–135). Boston: Martinus Njihoff.

Mikosch, T. (2005). How to model multivariate extremes if one must. *Statistica Neerlandica, 59*(3), 324–338.

Mitzenmacher, M. (2003). A brief history of generative models for power law and log normal distributions. *Internet Mathematics, 1*(2), 226–251.

Molteno, T. C. A. (1993). Fast $O(N)$ box-counting algorithm for estimating dimensions. *Physical Review E, 48*(5), R3263–R3266.

Mulligan, G. F. (1984). Agglomeration and central place theory: A review of the literature. *International Regional Science Review, 9*(1), 1–42.

Nams, V. O. (2006). Improving accuracy and precision in estimating fractal dimension of animal movement paths. *Acta Biotheoretica, 54*(1), 1–11.

Pareto, V. (1897). The new theories of economics. *Journal of Political Economy, 5*(4), 485–502.

Pastor-Satorras, R., Vazquez, A., & Vespignani, A. (2001). Dynamical and correlation properties of the Internet. *Physical Review Letters, 87,* 258701.

Peitgen, H.-O., Jürgens, H., & Saupe, D. (1992). *Chaos and fractals: New frontiers of science.* New York: Springer-Verlag.

Peng, C.-K., Buldyrev, S. V., Havlin, S., Simons, M., Stanley, H. E., & Goldberger, A. L. (1994). Mosaic organization of DNA nucleotides. *Physical Review E, 49*(2), 1685–1689.

Peng, S.-K. (2004). Advanced insights in central place theory. In R. Capello & P. Nijkamp (Eds.), *Urban dynamics and growth: Advances in urban economics* (pp. 413–442). Amsterdam: Elsevier.

Rice, J. A. (1988). *Mathematical statistics and data analysis*. Pacific Grove, CA: Wadsworth & Brooks/Cole Advanced Books & Software.

Rice-Snow, S., & Emert, J. W. (2002). A nonstratified model for natural fractal curves. *Mathematical Geology, 34*(5), 543–553.

Richardson, L. F. (1941). Frequency of occurrence of wars and other fatal quarrels. *Nature, 148*, 598.

Richardson, L. F. (1948). Variation of the frequency of fatal quarrels with magnitude. *Journal of the American Statistical Association, 43*(244), 523–546.

Richardson, L. F. (1960). *Statistics of deadly quarrels.* (Q. Wright & C. C. Lienau, Eds.). Pittsburgh, PA: Boxwood Press.

Roberts, D. C., & Turcotte, D. L. (1998). Fractality and self-organized criticality of wars. *Fractals, 6*(4), 351–357.

Rodin, V., & Rodina, E. (2000). The fractal dimension of Tokyo's streets. *Fractals, 8*(4), 413–418.

Rodkin, M. V., & Pisarenko, V. F. (2008). Damage from natural disasters: Fast growth of losses or stable ratio. *Russian Journal of Earth Sciences,* 10 ES1004 (doi:10.2205/ 2007ES000267)

Sambrook, R. C., & Voss, R. F. (2001). Fractal analysis of US settlement patterns. *Fractals, 9*(3), 241–250.

Sarraillé, J. J., & Myers, L. S. (1994). FD3: A program for measuring fractal dimension. *Educational and Psychological Measurement, 54*(1), 94–97.

Shen, G. (2002). Fractal dimension and fractal growth of urbanized areas. *International Journal of Geographical Information Science, 16*, 419–437.

Sims, D. W., Southall, E. J., Humphries, N. E., Hays, G. C., Bradshaw, C. J. A., Pitchford, J. W., James, A., Ahmed, M. Z., Brierley, A. S., Hindell, M. A., Morritt, D., Musyl, M. K., Righton, D., Shepard, E. L. C., Wearmouth, V. J., Wilson, R. P., Witt, M. J., & Metcalfe, J. D. (2008). Scaling laws of marine predator search behavior. *Nature, 451*, 1098–1103.

South, R., & Boots, B. (1999). Relaxing the nearest centre assumption in central place theory. *Papers in Regional Science, 78*, 157–177.

Spelman, W. (1994). *Criminal incapacitation.* New York and London: Plenum.

Spiegel, M. R., & Stephens, L. J. (2008). *Statistics* (4th ed.). New York: Schaum's Outline Series, McGraw-Hill.

Stanley, H. E., Amaral, L. A. N., Buldyrev, S. V., Gopikrishnan, P., Plerou, V., & Salinger, M. A. (2002). Self-organized complexity in economics and finance. *Proceedings of the National Academy of Sciences, 99*(Suppl. 1), 2561–2565.

Tanner, B. R., Perfect, E., & Kelley, J. T. (2006). Fractal analysis of Maine's glaciated shoreline tests established coastal classification scheme. *Journal of Coastal Research, 22*(5), 1300–1304.

Taylor, R. P., Micolich, A. P., & Jonas, D. (1999). Fractal analysis of Pollock's drip paintings. *Nature, 399*, 422.

Taylor, R. P., Micolich, A. P., & Jonas, D. (2002). The construction of Jackson Pollock's fractal drip paintings. *Leonardo, 35*(2), 203–207.

Turcotte, D. L. (1997). *Fractals and chaos in geology and geophysics* (2nd ed.). Cambridge, UK: Cambridge University Press.

Turcotte, D. L., & Huang, J. (1995). Fractal distributions in geology, scale invariance, and deterministic chaos. In C. C. Barton & P. R. La Pointe (Eds.), *Fractals in the earth sciences* (pp. 1–40). New York: Plenum.

Turcotte, D. L., & Malamud, B. D. (2004). Landslides, forest fires, and earthquakes: Examples of self-organized critical behavior. *Physica A, 340*, 580–589.

84

Turcotte, D. L., Pelletier, J. D., & Newman, W. I. (1998). Networks with side branching in biology. *Journal of theoretical biology 193*, pp. 577–592.

Turcotte, D. L., & Rundle, J. B. (2002). Self-organized complexity in the physical, biological, and social sciences. *Proceedings of the National Academy of Sciences, 99*(Suppl. 1), 2463–2465.

U.S. Census Bureau. (2006). *The 2006 statistical abstract of the United States: The national data book: Table 358: Oil Spills in U.S. Water—Number and Volume by Spill Characteristics: 1998 to 2001*. Retrieved June 19, 2008, from http://www.census.gov/compendia/statab/2006/tables/06s0358.xls

Verma, A. (1998). The fractal dimension of policing. *Journal of Criminal Justice, 26*(5), 425–435.

Viswanathan, G. M., Afanasyev, V., Buldyrev, S. V., Murphy, E. J., Prince, P. A., & Stanley, H. E. (1996). Lévy flight search patterns of wandering albatrosses. *Nature, 381*, 413–415.

Viswanathan, G. M., Buldyrev, S. V., Havlin, S., da Luz, M. G. E., Raposo, E. P., & Stanley, H. E. (1999). Optimizing the success of random searches. *Nature, 401*, 911–914.

Voit, J. (2003). *The statistical mechanics of financial markets* (2nd ed.). Berlin: Springer.

Voss, R. F. (1998). Local connected fractal dimension analysis of early Chinese landscape paintings and x-ray mammograms. In Y. Fisher (Ed.), *Fractal image coding and analysis* (pp. 279–297). Berlin: Springer.

Wen, R., & Sinding-Larsen, R. (1997). Uncertainty in fractal dimension estimated from power spectra and variograms. *Mathematical Geology, 29*(6), 727–753.

White, D. & Johansen, U. (2005). *Network analysis and ethnographic problems: Process models of a Turkish nomad clad*. Lanham, MD: Lexington Books.

Wichmann, S. (2005). On the power-law distribution of language family sizes. *Journal of Linguistics, 41*, 117–131.

Wong, D. W. S., Lasus, H., & Falk, R. F. (1999). Exploring the variability of segregation index *D* with scale and zonal systems: An analysis of thirty US cities. *Environment and Planning A, 31*, 507–522.

Zhou, W.-X., Sornette, D., Hill, R. A., & Dunbar, R. I. M. (2005). Discrete hierarchical organization of social group sizes. *Proceedings of the Royal Society B, 272*, 439–444.

Zipf, G. K. (1949). *Human behavior and the principle of least effort: An introduction to human ecology*. Cambridge, MA: Addison-Wesley.

AUTHOR INDEX

86

SUBJECT INDEX

Supporting researchers for more than 40 years

Research methods have always been at the core of SAGE's publishing program. Founder Sara Miller McCune published SAGE's first methods book, *Public Policy Evaluation*, in 1970. Soon after, she launched the *Quantitative Applications in the Social Sciences* series—affectionately known as the "little green books."

Always at the forefront of developing and supporting new approaches in methods, SAGE published early groundbreaking texts and journals in the fields of qualitative methods and evaluation.

Today, more than 40 years and two million little green books later, SAGE continues to push the boundaries with a growing list of more than 1,200 research methods books, journals, and reference works across the social, behavioral, and health sciences. Its imprints—Pine Forge Press, home of innovative textbooks in sociology, and Corwin, publisher of PreK–12 resources for teachers and administrators—broaden SAGE's range of offerings in methods. SAGE further extended its impact in 2008 when it acquired CQ Press and its best-selling and highly respected political science research methods list.

From qualitative, quantitative, and mixed methods to evaluation, SAGE is the essential resource for academics and practitioners looking for the latest methods by leading scholars.

For more information, visit **www.sagepub.com**.

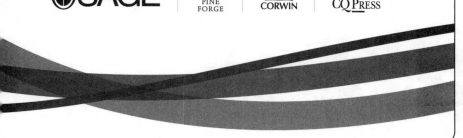